HOW TO REMOVE A BRAIN

HOW TO REMOVE A BRAIN

AND OTHER BIZARRE MEDICAL PRACTICES AND PROCEDURES

DAVID HAVILAND

THISTLE
PUBLISHING

First published in 2012 by Summersdale Publishing

This edition published in 2017 by:

Thistle Publishing
36 Great Smith Street
London
SW1P 3BU

www.thistlepublishing.co.uk

CHAPTER ONE
THE WISDOM OF THE ANCIENTS

*'Physicians of all men are most happy;
what good success soever they have the world proclaimeth,
and what faults they commit, the earth covereth.'*

Francis Quarles

What unusual use did ancient Egyptians find for crocodile dung?

Bizarrely, crocodile dung was used as a contraceptive device in ancient Egypt. While the ancient Egyptians had a fairly complex and advanced system of medicine involving herbal drugs, poultices, laxatives, suppositories, surgery, bonesetting, ophthalmology, and even a system of health insurance, the vast majority of Egyptian treatments were totally ineffective, and sometimes even quite harmful. For example, one Egyptian cure for impotence contained 39 separate exotic ingredients, none of which would have had any useful effect.

Using crocodile dung as a method of contraception may sound pretty daft, but it was probably effective to some degree. The dried dung was used as a pessary, which would be inserted into the vagina. The idea was that it would soften as it reached body temperature, and thus form a secure, impenetrable barrier against the cervix. Cervical caps of this type are used as contraception today, although thankfully they tend not to be made of dung. Furthermore, the acidity of the crocodile dung would probably have acted as a mild spermicide, offering some additional protection.

Nonetheless, putting crocodile dung into your body in any form is not to be recommended. Dung is full of bacteria, parasites, and other germs, so there is a considerable risk of infection. It's also just plain gross. Other traditional pessaries down the ages have been made from elephant dung, tree sap, halves of lemon, cotton, wool, and natural sea sponges, and each of them was probably only about as effective as the others.

Which society believed dead mouse paste could cure toothache?

In ancient Egypt, one recommended cure for toothache was to apply a dead mouse to the tooth or gum. Alternatively, the patient could mash the mouse into a paste, and mix it with other ingredients before applying it.

The ancient Egyptians weren't alone is extolling the benefits of mouse poultices. In Elizabethan England, one cure for warts was to cut a mouse in half, and apply it to the offending pustule. The Elizabethans also ate mice, either fried, or baked in pies. As well as curing warts, mice were also believed to remedy whooping cough, measles, smallpox, and bed-wetting.

How do you remove a brain?

From around 3500 BC onward, the ancient Egyptians developed a complex system of mummification, preserving the corpses of their dead by drying them out, removing the internal organs, and wrapping the corpse in bandages. This practice may well have been inspired by the natural mummification which took place when bodies were buried in the arid Egyptian desert. Having observed that a body could be perfectly preserved after death, the Egyptians seem to have developed a belief that preservation of the body was necessary, if the spirit was to survive in the afterlife. They believed that the soul consisted of three separate spirits, and one of these, Ka, was intimately bound up with the physical body. Unless the body was preserved in this world, the Ka could not survive in the next.

The mummification process was a complex, painstaking ritual. The body would be taken to Ibu, the 'place of purification', and washed in the waters of the Nile. It would then be taken to Per-Nefer, the 'house of mummification', to be embalmed. First, the brain would be removed and discarded, as it was thought to be unimportant. Then, a small slit was cut along the left side of the body, through which the internal organs were removed, to prevent the body from decomposing from the inside. The kidneys would be discarded, presumably because the Egyptians thought they served no useful purpose. The heart was left in the body, as it was considered to be the centre of a person's being. The rest of the organs would be kept in jars, and placed in the coffin, as the Egyptians believed the reincarnated spirit would need them in the next life. For the same reason, the wealthy would be buried with their important possessions, jewellery, precious metals, sacred charms, amulets, books of spells, furniture, clothing, food, and even their mummified pets. The corpse would then be stuffed with incense and other material, to return it to a life-like shape. The body would then be completely covered with salty 'natron' powder, for around 35–40 days, to dry it out, after which it would be restuffed, and then carefully wrapped in bandages.

When European explorers began to take a fresh interest in the Egyptians' lavish tombs in the 19th century, they were faced with a number of mysteries. One of these was the question of how the Egyptians had managed to remove the brains of their dead. There was no evidence of damage to the mummies' craniums, and yet the brains had been completely removed. How was this possible? The answer was simple: they went in through the nose. The embalmers would use a long wire with a hook on the end, which they would force through the nose, to scrape out the brain, chunk by chunk. Once all the brain matter had been removed, the inside of the skull would be washed, again via the nasal cavity.

Bizarrely, it seems that a similar technique may well end up having a dramatic effect on modern-day brain surgery. Until recently, the usual procedure for removing a brain tumour at the base of the skull was to remove either part of the skull, or part of the facial skeleton. Either method would cause dramatic blood loss and risk of infection, as well as considerable discomfort and scarring. However a new procedure, which is called Endoscopic Transnasal Brain Surgery, involves inserting an endoscope through the nose, and guiding it directly to the site of the tumour. The endoscope contains a tiny video camera, which transmits live images into the operating theatre. The endoscope also incorporates specially designed surgical tools, which can be used to dissect and remove the tumour. Because this new technique is far less damaging, it can reduce the recovery time after surgery to a matter of days, whereas conventional procedures take weeks or even months to heal.

When was plastic surgery invented?

Surprisingly perhaps, plastic surgery has been around for more than 3,000 years. In India, records have been found detailing ancient procedures for repairing a broken nose, and suturing to avoid scarring. Around 500 BC Hindu doctor named Susrata developed a procedure to repair noses which had been cut off as a punishment for adultery (for some reason, it was determined that the interloper in the marriage was most at fault, so it was this third party who would lose their nose). Susrata found a way to repair this shameful injury, which was of course intended to cause the maximum social stigma, by taking skin from the cheek or forehead.

In 1597 Italian doctor Gasparo Tagliacozzi improved the procedure by lifting a skin flap from the arm and stitching it on to the nose, while it was still simultaneously attached to the arm. Once the graft had taken hold, the flap would be cut free from the arm. There was considerable demand for nose surgery in Europe from the 15th century onwards, due to the dreadful effects of syphilis, which could cause sufferers to lose their nose.

However, before the development of effective anaesthetics in the 1840s, any kind of surgery was incredibly painful, not to mention dangerous, so the idea of plastic surgery for purely cosmetic reasons was unthinkable. Interestingly, the word 'plastic' in the context of plastic surgery doesn't mean artificial, and nor does it refer to the materials used. Instead, it comes from the Greek word 'plastikos' which means to mould or shape, in the same sense that ceramics and sculpture are known as the 'plastic arts'.

Do doctors give an oath vowing never to perform euthanasia, abortion, or surgery of any kind?

Hippocrates (c. 460–377 BC) is regarded as the father of modern medicine. He was born on the island of Cos, and he and his followers published a great number of medical texts, which formed the foundation of Western medicine until the Enlightenment. Hippocrates was the first doctor to reject the prevailing superstitious belief that illness was caused by the gods. Instead, Hippocrates argued that illness was the product of the patient's environment, diet, and lifestyle, and that therefore professional physicians could bring about natural means of healing, without requiring the intervention of the gods. The Hippocratic school left behind around 60 works, which are known as the Hippocratic *corpus*, and which formed the basis of medicine until the nineteenth century.

To address concerns about the ethics of medical practice, Hippocrates and his followers produced a detailed oath, to demonstrate the physician's devotion to his art, and to the patient. It is in this oath that the first principle of Hippocratic medicine is outlined: 'primum non nocere', which means 'to do no harm'. A version of this oath is still taken by the majority of doctors today. However, when people refer to the Hippocratic Oath nowadays, they often mean the principle that anything told to the doctor must be treated as confidential, and it is true that this is one of the principles of the oath. However, the oath also contains a number of other points. Here is the original oath in full:

I swear by Apollo the healer, by Aesculapius, by Health and all the powers of healing, and call to witness all the gods and goddesses that I may keep this Oath and Promise to the best of my ability and judgement.

I will pay the same respect to my master in the Science as to my parents and share my life with him and pay all my debts to him. I will regard his sons as my brothers and teach them the Science, if they desire to learn it, without fee or contract. I will hand on precepts, lectures and all other learning to my sons, to those of my master and to those pupils duly apprenticed and sworn, and to none other.

I will use my power to help the sick to the best of my ability and judgement; I will abstain from harming or wronging any man by it.

I will not give a fatal draught to anyone if I am asked, nor will I suggest any such thing. Neither will I give a woman means to procure an abortion.

I will be chaste and religious in my life and in my practice.

I will not cut, even for the stone, but I will leave such procedures to the practitioners of that craft.

Whenever I go into a house, I will go to help the sick and never with the intention of doing harm or injury. I will not abuse my position to indulge in sexual contacts with the bodies of women or of men, whether they be free-men or slaves.

Whatever I see or hear, professionally or privately, which ought not to be divulged, I will keep secret and tell no one.

If, therefore, I observe this Oath and do not violate it, may I prosper both in my life and in my profession, earning good repute among all men for my time. If I transgress and forswear this oath, may my lot be otherwise.

(Translated by J Chadwick and WN Mann, 1950)

The highlighted lines do seem to state that takers of the oath will not perform euthanasia, abortion, or surgery, so how do today's doctors reconcile this oath with modern medical practices? The simple answer is that they take an amended version of the oath. The most common oaths taken today are the Declaration of Geneva, the Prayer of Maimonides, the Oath of Lasagna, and the Reinstatement of Hippocratic Oath. All four are based on the Hippocratic Oath, and contain many of the same vows, but each is amended to be more appropriate for modern ethics and medical practices.

Most graduating medical students today take one of these four versions of the Hippocratic Oath before going out into the world to practise medicine. In a survey of the oaths taken in 150 US and Canadian medical schools in 1993, it was found that only 14% forbade euthanasia, 8% prohibited abortion, and only 3%

banned sexual contact with patients. The line referring to surgery is usually retained in some form, but in today's oaths it is treated as a metaphor, the purpose of which is to acknowledge that no doctor can maintain expertise in all fields.

Where do maggots come from?

Maggots are the larvae of flies. They emerge from eggs, which have usually been laid directly onto the kind of thing that flies consider to be food, such as rotting animal corpses. However, for centuries people believed that maggots would simply emerge from a corpse, in a process called 'spontaneous generation'. In other words, people believed that living things could be generated from non-living things, without the presence of an egg, larva, or parent. Spontaneous generation was the accepted explanation for the origins of many types of creature for over 2,000 years.

The theory of spontaneous generation was notably advanced by Aristotle (384 BC – 322 BC) in his text 'The History of Animals'. Aristotle believed that fleas and maggots emerged spontaneously from rotting flesh, that mice were created by piles of dirty hay, and aphids would magically arise from the morning dew. Of course, some creatures were obviously produced by parents, and Aristotle recognised this, but he thought there was also another class of creatures, which could emerge from non-living things. Astonishingly, this theory of spontaneous generation held sway until as late as the 19th century.

However, with the development of increasingly powerful microscopes, scientists began to observe smaller and smaller life forms, which once again raised the question of generation – were these tiny micro-organisms proof of spontaneous generation, or did they show that there was a whole world of tiny lifeforms about which we knew nothing? A Dutch scientist named Anthony van Leeuwenhoek observed tiny creatures in the microscopes he expertly built for himself, and he called these tiny lifeforms 'animalcules'. Van Leeuwenhoek noticed that living micro-organisms would appear in rainwater after just a few days, which raised the question of where they had come from.

In the 18th and 19th centuries, a series of experiments would conclusively answer this question, and in the process they would pave the way for modern, effective medicine. In 1748, John Needham performed a series of experiments on jars of meat broth. Needham believed that boiling the broth would kill any living animalcules, and so when the broths turned cloudy soon afterwards, indicating that they gone off, this confirmed his theory that the animalcules must have spontaneously emerged from the broth. In 1765, Lazzaro Spallanzani carried out a similar experiment, but unlike Needham he sealed the lids of the jars, making them airtight, to prevent any contaminants from getting into the jars after boiling. Because the jars had been sealed, the broth did not produce any microbes, which seemed to disprove the theory of spontaneous generation. However, suuporters of the theory argued that no air had been allowed to reach the broth, and that air was one of the elements which was crucial in somehow facilitating spontaneous generation.

In 1859, Louis Pasteur resolved this difficult impasse, by boiling broth in a specially designed swan-necked flask, which had an extremely long, thin neck which curved downwards. This neck would allow air to reach the broth, but no micro-organisms could navigate the narrow bend. After boiling, the broth in the swan-necked flasks remained free of microbial growth, demonstrating that air alone wasn't sufficient to generate micro-organisms. This elegant experiment, which could be easily repeated and tested by anyone, provided a powerful refutation of the theory of spontaneous generation, and was thus a key moment in the development of what became known as 'germ theory'.

How was Julius Caesar born?

There is an enduring myth that the Roman leader Julius Caesar was born by Caesarian section, and that consequently the operation was named after him, or alternatively that he was named after the operation. This story is widely known, and has even made it into the Oxford English Dictionary. Nonetheless, it is almost certainly false.

Firstly, what exactly is a Caesarean section? It is an operation in which an expectant mother's abdomen is cut open to deliver the baby, when a vaginal delivery is thought to be too dangerous. At the time of Julius Caesar's birth, and for centuries afterwards, a Caesarian birth was an extreme measure, which would only be carried out if the mother had died, or was in fatal danger. In Roman times, there was no effective anaesthesia, and doctors had no proper way of suturing the enormous wounds which constitute the operation. Even on those rare occasions when doctors did manage to stem the haemorrhaging, the wounds were pretty much certain to become infected.

Caesarean births are known to have predated Julius Caesar, but there are no records of any mother surviving a Caesarean birth until 1500 AD. This is one reason we can be sure Caesar himself was not born by this method, as his mother didn't die until he was 46. Also, Julius Caesar couldn't have been named after the operation, because the name 'Caesar' had already been in his family for generations. The dynasty kept detailed ancestral records, and claimed that they could trace their lineage back to the Trojan prince Aeneas and the goddess Venus.

However, although Julius Caesar himself was not born by Caesarean, there does seem to be a link between the two names. The name 'Caesar' may have entered the family because one of Julius Caesar's distant ancestors was born by this method, according to the Roman writer Pliny the Elder. The name may therefore have come from the Latin word 'caedare', meaning 'to cut', or alternatively from a law called 'Lex Caesarea', which required the operation to take place if the mother died during childbirth.

On the other hand, the family name 'Caesar' may have had nothing to do with Caesarean sections. Instead, it may have referred to an ancestor who was born with a full head of hair, from 'caesaries' meaning 'hair'; or an ancestor who had blue-grey eyes, from 'caesius' meaning 'blue-grey'; or from 'caesai', the Punic word for 'elephant', in reference to an ancestor who had killed an elephant during the First Punic War. During his reign, Julius Caesar had coins made with an image of an elephant above the name 'Caesar', which suggests that this was his preferred theory.

What was the 'Doctrine of Signatures'?

The Doctrine of Signatures was an ancient diagnostic system, which was based upon the simple principle that God marked everything he created with a sign, and that this sign was a clue as to each thing's purpose. Therefore, walnuts were believed to be good for the brain, because they look a bit like brains, as well as being, like brains, a fairly soft, spongey substance enclosed within a hard shell. The Doctrine of Signatures was a central plank of medical theory going back at least as far as Galen of Pergamum (129–200 AD), who was the most celebrated doctor of the Roman period, and the Doctrine remained central to medical thinking until as recently as the late 19th century.

According to the Doctrine of Signatures, natural remedies embody or reflect the ailments they are designed to cure. Thus, lungwort was believed to be good for lung conditions, because its white-spotted, oval leaves look a bit like diseased lungs. Cardamine flowers, which are also known as toothwort, were meant to be good for toothache, as these small, white flowers have a similar shape to teeth. St John's Wort was said to be good for the skin, because the oil glands in the plant's leaves look like skin pores. In a few cases, there was some truth to these claims – St John's Wort, for example, is recognised by modern medicine to contain a strong antibiotic, which helps wounds to heal quickly (incidentally, St John's Wort is also used in herbal medicine as an anti-depressant, although recent studies seem to be inconclusive as to whether or not it has anything more than a placebo effect).

More often, though, the remedies recommended by the Doctrine had no real medicinal value. For example, henbane was recommended for toothache, because its seed container is shaped like a human jaw, but henbane is actually a poisonous hallucinogen, which is potentially fatal. Nowadays, of course, the Doctrine of Signatures is not taken seriously by mainstream medicine, and any successes it might claim are regarded as simply coincidences, or 'attribution after the fact'. In other words, when people noticed that St John's Wort helped wounds to heal, they then found a tenuous way to argue that it resembles skin, to fit in with their contemporary understanding of medicine.

However, there are two growing areas of modern healthcare in which the Doctrine of Signatures is still taken seriously. Firstly, the doctrine is used by modern herbalists, either for clues to help them rediscover forgotten essences, or to support the claims made for existing herbal products. Secondly, the Doctrine is one of the two defining principles of homeopathy, in which context it is often summarised as 'like cures like'. The other key homeopathic principle is the 'law of infinitesimals', which holds that the smaller the dose, the more effective the medicine. In accordance with this second law, homeopathic remedies are diluted to such an enormous extent that they actually contain little else but water, although advocates believe that this water has somehow retained a memory or imprint of the original substance. Despite being dismissed as quackery by mainstream science, over the last few years homeopathy hass been increasingly funded and supported by Britain's National Health Service.

Were the Incas the first to carry out blood transfusions?

In the history of European medicine, the key development in the story of blood transfusions was Karl Landsteiner's discovery in 1901 that different people have different blood types. Before this point, blood transfusions had been tried many times, but they were often fatal, because our bodies undergo an immunological reaction to blood of a different type. Landsteiner correctly concluded that it was necessary to match the blood type of the donor to that of the recipient.

Before this discovery, many types of blood transfusion had been attempted. In 1492, Pope Innocent VIII was given the blood of three young boys after slipping into a coma, but he and the boys all died as a result. In 1667, French doctor Jean-Baptiste Denys gave eight ounces of lambs' blood to a young man with a fever, who had been repeatedly bled (at this time, bloodletting was a common medical treatment, used for almost any ailment), and the man survived. Denys also performed a transfusion of lambs' blood into a labourer, who also survived. In both cases, the amount of blood transfused was relatively small, which presumably explains how the patients managed to survive the allergic reaction. Denys's next patient died, and as a result blood transfusions were banned in both France and Britain. However, rumours circulated that the unfortunate patient may have died not from the transfusion, but from poison administered by his own wife!

There are some reports which suggest that blood transfusions may have been successfully performed centuries before Denys, by the Inca people of South America. The Incan Empire rapidly expanded during the 15th and 16th centuries, covering much of the west coast of South America, from as far north as present day Colombia down to present day Argentina. Incan medicine was a bloody business. The best record of their civilisation was written by 'El Inca' Garcilaso, who was the product of the relationship between a Spanish conquistador and an Incan princess. In his writing, El Inca described a system of medicine based largely around bloodletting and purging. Patients would be bled from the arm or leg nearest to the site of the ailment. A headache, for example, would be treated by bloodletting from the forehead, at the spot where the eyebrows meet.

The reason why the Incas managed to succeed with blood transfusions, while European medicine had until that point failed, may simply be that they all had the same blood type. It seems that a vast majority of Incas were of blood type O, rhesus positive, which means that there would have been very few incompatibility reactions.

CHAPTER TWO
DISGUSTING DISEASES

'Where the sun enters, the doctor does not'
Traditional proverb

Are there men who have to wheel their testicles around in a wheelbarrow?

Amazingly, some sufferers of elephantiasis really do have to do just this, because their scrotum has swollen to such an enormous size. Elephantiasis is an unpleasant, disfiguring disease, which is found in many parts of Africa, India and South Asia, and affects more than 120 million people worldwide. In communities where the disease is endemic, it can affect between 10% and 50% of men, and around 10% of women.

Elephantiasis is caused by long, thread-like filarial worms, which are spread by mosquitoes, and reproduce in the human bloodstream, infecting the lymphatic system. Infected people often remain symptomless for years, while the worms rapidly multiply in their blood. When symptoms do appear, they are dramatic. Sufferers develop grotesquely swollen limbs, genitals, and breasts. Men are more likely to be victims than women, and swelling of the penis and scrotum is common.

Recent developments have made elephantiasis easier to diagnose in symptomless sufferers, and the disease can be treated with antibiotics. Surgery can also be helpful for victims of scrotal elephantiasis. The World Health Organisation is working towards an ambitious target to completely eliminate the disease by 2020, and so far it says it's on track.

So does this mean there will never be another Elephant Man?

Actually, although Joseph Merrick - the man who came to be known as the Elephant Man - was once thought to have suffered from elephantiasis, it now seems more likely that he was in fact suffering from a combination of two rare disorders: Proteus Syndrome and Neurofibromatosis type I.

He was born in Leicester in 1862, with no visible signs of any unusual condition. Merrick was a healthy, normal child until around the age of five, when his skin began to change. It became thick, lumpy, and grey, and he developed a swelling on his upper lip. As he grew older, his right arm became grotesquely swollen, while his left arm remained thin but normal. The swellings on his face continued to grow, along with a large, bony lump on his forehead. At one point he fell and damaged his hip, leaving him permanently lame.

After leaving school, he took on a number of jobs, but his deteriorating condition and shocking appearance made work impossible. He worked in a factory rolling cigars, until his fingers became too thick, losing the dexterity and nimbleness required for the job. He then worked, astonishingly, as a door-to-door salesman, but potential customers were shocked by his appearance, and by now the swellings on his face made his speech incomprehensible. He ended up in a workhouse, before deciding that his only option was to join a freak show.

Merrick teamed up with a promoter, who exhibited him to the public as 'The Elephant Man', telling a fantastical story about how Merrick's mother had been knocked down by a rampaging elephant during pregnancy, which had caused the infant to be born disfigured. This story sounds like a showman's absurd invention, but in fact there was an elephant on the loose in Leicester in 1862, so there may be an element of truth to the story (although even if Merrick's mother had been knocked over by an elephant, it wouldn't have caused his condition).

One of the visitors to the freak show was a doctor named Frederick Treves, who took a particular interest in Merrick, and examined him at his hospital on a number of occasions. After a disastrous attempt to take the show to Europe, Merrick was robbed and left destitute, and so Treves took him in, admitting him to live at the London Hospital. Treves then organised a public campaign to raise funds for his care, and as a result Merrick became famous. Over time, he came to be looked upon with a great deal of public sympathy, and was befriended by many of the great and good, including Princess Alexandra and Queen Victoria.

Apart from the exact medical nature of his condition, there remains another enduring mystery concerning Merrick: how did he die? David Lynch's film 'The Elephant Man' shows Merrick suffocating in his sleep, but this is not what is actually believed to have happened. When his body was found, Merrick had dislocated his neck, and it was this which killed him. Merrick's head was too heavy for him to sleep in a reclining position, but it's possible that on this particular night he may have wanted to forget his disfigurements, and to lie down to sleep just once like a normal person.

Despite the unimaginable hardships he faced, Merrick was a gentle and cultured man. He often ended his correspondence with the following poem by Isaac Watts:

'Tis true my form is something odd,
But blaming me is blaming God.

Could I create myself anew,
I would not fail in pleasing you.
If I could reach from pole to pole,
Or grasp the ocean with a span,
I would be measured by the soul,
The mind's the standard of the man.

Today, there are effective treatments and surgerical procedures for both Proteus Syndrome and Neurofibromatosis type I. Since the symptoms of someone like Joseph Merrick are so visibly dramatic, it seems unlikely that there will ever be another Elephant Man, as today anyone exhibiting such symptoms would surely receive quick and effective treatment.

How often can a person vomit?

Perhaps more often than you might think. There is a particular condition called Cyclic Vomiting Syndrome (CVS) in which sufferers regularly vomit as often as twelve times per hour - that's every five minutes - for weeks on end. In between these periods, there will be intervals with no symptoms, and there is usually a set pattern of episodes. This unpleasant condition was first described in 1882 by doctor Samuel Gee. It tends to develop in children between the age of 3 and 7, and it can continue into adulthood. Many sufferers find the condition too debilitating to go to school or work, while others seem to be less dramatically affected.

It's not clear what causes CVS, but it is thought to be somehow linked to migraine headaches. Many CVS sufferers also have a family history of migraine, and the two conditions are quite similar in some ways, as they both consist of sudden, intense symptoms, some of which overlap, interspersed with pain-free intervals, and both conditions are triggered by many of the same stimuli. Some CVS sufferers say that their 'triggers' are very predictable, and they can include illness, stress, certain foods, excitement, anxiety, and panic attacks.

Rest and sleep are normally recommended when a CVS bout begins, and many sufferers learn to physically control their symptoms. Frequent vomiting can cause dehydration and a loss of electrolytes, so sufferers are sometimes given painkillers or anti-nausea medicine. Vomiting can also irritate or injure the oesophagus because of the stomach acid, so vomit will sometimes contain blood and/or bile.

What is The King's Evil?

This was the colloquial name given to a horrible disease called 'scrofula', which is a tuberculous infection of the skin of the neck. Scrofula affects the lymph nodes, causing large blue or purple swellings to grow on the neck and chest. These lumps are initially painless, if unsightly, but after a certain time they burst, leaving a nasty open wound. Scrofula is usually caused by the *Mycobacterium tuberculosis* bacterium in adults, which can be passed from person to person through breathing. As tuberculosis declined during the 20th century, scrofula was almost wiped out, but with the advent of AIDS-weakened immune systems, it is becoming more common again.

Today, most types of scrofula can be easily treated with antibiotics, but in past centuries there was no cure or effective treatment. The disease became known as 'The King's Evil', because it was thought that a touch from the king would cure the swellings. This royal healing power was said to have been passed down from Edward The Confessor. Many of the kings of England and France would indulge in this practice, touching huge numbers of victims, and handing out coins to the afflicted. Henry IV of France is reported to have touched as many as 1,500 scrofula victims during one sitting. Charles II of England is recorded as bestowing the royal touch on 92,107 of his subjects over the course of his reign, although quite why such detailed records were kept is not clear. The great man of letters Dr Johnson was touched by Queen Anne for scrofula when he was a child. The practice died out in the 18th century, as it was felt, in England at least, to be too Catholic. It also suffered a crucial blow in France when the Dauphin Louis XVII died of scrofula, which must have rather undermined the theory.

Why did people believe that keeping farts in a jar could ward off the Black Death?

The Black Death was one of the deadliest plagues in human history. Between 1348 and 1350 it killed around 1.5 million people in Britain, out of a population of just 4 million. The plague was characterised by unpleasant black buboes which would appear in the victim's groin, neck and armpits, oozing pus and blood. The disease caused fever, nausea, and vomiting, and most victims died within four to seven days after becoming infected. The plague began to recede in England after 1350, but it never completely went away, and there were regular outbreaks about once every generation, until the last major outbreak in 1665, the Great Plague of London.

Medicine at the time was based upon principles which had been around since the ancient Greeks, including the simple tenet that 'like cures like', also known as the Doctrine of Signatures. Since the plague was believed to be caused by deadly vapours, it therefore seemed to make sense that other foul smells might help to ward off the disease. Some doctors therefore recommended keeping dirty goats inside the home, to create a therapeutic stink. Others suggested using another source of foul odours: our own farts. However, rather than wasting the precious pong, people were instead advised to store their farts in jars, which could then be opened and inhaled the next time the deadly pestilence appeared in the neighbourhood.

Which disease is called 'trembling with fear'?

The answer is kuru, a fascinating, incurable brain disease which, so far at least, has only ever affected a tiny tribe known as the Fore (pronounced for-ay) people, who are found in the highlands of New Guinea. The word 'kuru' means 'trembling with fear' in the Fore's native language, and it refers both to the physical tremors that are a key symptom of the disease, and its dreadful fatality rate. After the first onset of symptoms, kuru victims are practically certain to die within a year or two, a fatality rate which makes kuru one of the deadliest diseases known to man.

Kuru was first discovered in 1957, and at that point it may not have even existed for very long. According to the older members of the Fore tribe, the disease had not been around when they were young, which suggests it couldn't have existed for much more than twenty years. The Fore people were a completely isolated community, thanks to the mountainous terrain of New Guinea, and there are no reports of kuru ever occurring anywhere else.

The disease is believed to have spread as a result of the Fore's cannibalistic funeral rites. When a member of the tribe died, the female relatives would ritualistically dismember the corpse, removing the arms and feet, stripping the muscle from the body, cutting open the chest to remove the internal organs, and scooping out the brain. The tribe would then cook and eat the corpse, including the brain, which is thought to be the most infectious organ.

Meat from the bodies of kuru victims was particularly prized, as the layers of fat on those who had died apparently resembled pork. The men of the tribe would be given the best cuts of meat, leaving the rest of the body, including the brain, to the women and children, which explains why kuru was around 8–9 times more prevalent among women than among men. An alternative explanation for this fact is that it was the women who were responsible for dismembering the bodies, and they may therefore have become infected via open sores and cuts coming into contact with the disease, rather than because of ingesting the brain material.

Kuru quickly developed into an epidemic, with dramatic effects. The first symptoms of kuru are headaches, joint pain, physical tremors, and a gradual loss of limb control. Some victims will also burst into pathological fits of laughter. Over time, sufferers become unable to stand, and eventually unable to eat. As a result, many tribespeople died of starvation during the epidemic. Between 1957 and 1968, more than 1,100 people died of kuru, out of a population of just 8,000. There is no cure or treatment, but the disease gradually died out, as the efforts of the Australian government and Christian missionaries eventually persuaded the Fore to stop carrying out their cannibalistic death rituals.

Kuru is an extraordinary disease in one other respect. It is an infectious disease, but it is not caused by a virus, bacterium, or a parasite. Instead, it is caused by prions, which are mis-shapen proteins, which somehow cause other proteins in the body to lose their shape. Other prion diseases include Creutzfeldt-Jakob disease, BSE (mad cow disease), and scrapie, which are collectively known as transmissible spongiform encephalopathies (TSEs), which essentially means that they make the victims' brains become spongy and full of holes. Thankfully, kuru now seems to have been completely eliminated.

Why would the Ebola virus be a poor choice for biological warfare?

Ebola is a horrific disease, which destroys blood vessel cells, leading to massive internal bleeding. It is infectious, highly fatal, and endemic in parts of tropical Africa. The disease was first observed in Zaire in 1976, in an outbreak that infected more than 300 people, 90% of whom were killed.

However, although the Ebola virus is both deadly and contagious, it would actually make a poor choice as a biological weapon for one simple reason: it kills its victims too quickly. Viruses such as influenza can spread quickly over whole continents because they are easily transmitted, their initial symptoms are mild, and their victims are often contagious for a relatively long time.

Ebola, on the other hand, is not so easily transmitted, for a number of reasons. Firstly, it does not seem to be particularly contagious in its early stages. After an incubation period of around 5–18 days, symptoms will begin, and these can be severe, including fever, abdominal pain, and bloody vomit. After the onset of these symptoms, most patients have only around two weeks to live – most die of multiple organ failure or massive loss of blood. The 'case fatality rate' of some strains of Ebola may be as high as 90%, meaning that 90% of those infected die from the disease. By comparison, the case fatality rate for Spanish flu, the worst pandemic in human history, which killed around 50 million people between 1918 and 1920, was only around 5–10%. Ebola victims are contagious for such a short time, with such severe symptoms, that there is only limited scope for them to transmit the disease to others.

However, although Ebola does not seem to be an ideal choice for biological warfare, this doesn't mean people haven't tried. In 1992, a bizarre Japanese religious cult called Aum Shinkrikyo tried to acquire a sample of the Ebola virus for use as a biological weapon. Cult leader Shoko Asahara led a group of around 40 followers to Zaire, under the pretext of a medical mission, to try and get hold of a sample of the disease. The group were later convicted of the Sarin gas attack on the Tokyo subway in 1995, and when police raided the cult headquarters at the foot of Mount Fuji, they found a range of chemical and biological warfare agents, including anthrax and Ebola, as well as guns, explosives, and a Russian Mil Mi-17 military helicopter. The Sarin gas attack was the worst terrorist atrocity in Japanese history, and the group's motives remain unclear. One theory is that the attack was, bizarrely, an attempt to draw police attention away from an ongoing investigation into the group.

Can people grow horns?

Amazingly, they can. There is a rare condition called Cornu Cutaneum, in which sufferers grow strange, conical protrusions which are hard and brittle, resembling horn, wood, or coral. These horns, which are known as cutaneous horns, are usually found on visible parts of the body, such as the face, ears, forearms, and hands, which suggests that they may be linked to radiation from sun exposure, as these are the parts of the body that tend to see the sun. They may also be linked to the human papilloma virus family, which causes warts.

Cutaneous horns are tumours which can be benign, premalignant, or malignant, and surgery is usually the recommended treatment. They are made of keratin, which is a versatile protein found in our hair and fingernails, as well as in various animals' horns, hooves, and claws.

In one astonishing recent case, an elderly Chinese woman was found who appeared to be growing large, devil-like horns. On one side of her forehead, the 101-year-old Zhang Ruifang had grown a thick, black horn that was more than six centimetres long. Then, another matching horn started to grow on the other side of her forehead, making her look quite, well, devilish...

Do we share any diseases with animals?

Indeed we do. In fact, we share hundreds of diseases with animals, and they are known as 'zoonoses' (the singular is zoonosis). In fact, a great many of the diseases which affect humans today were first suffered by animals. At the end of the last Ice Age, around 10,000–12,000 years ago, humans began to move from a nomadic hunter-gatherer society to one based on settlements and agriculture. One consequence of this was that we began to live in close proximity to animals, including poultry, dogs, pigs, horses, sheep, and cattle.

At this point, illnesses which had previously evolved to infect animals began the process of mutating and leaping the 'species gap' to infect humans. From horses, we caught the common cold. From dogs, we caught measles. Pigs and ducks passed on their influenzas, while smallpox and tuberculosis were transmitted to us from cattle. Today, we share around 60 diseases with dogs alone, and only slightly fewer with pigs, goats, sheep, horses, and cattle.

Furthermore, this process is still ongoing, as diseases continue to transfer from animals to humans. HIV originated with primates, and only transferred to humans around the start of the 20th century. In 1994, a new virus called Hendra was discovered, after an outbreak which caused the death of 14 horses. In the second outbreak later that year, two more horses died, along with their owner. Even more recently, both avian flu and West Nile virus have crossed over from animals into human populations.

What was Chimney Sweep's Scrotum?

The occupation of a chimney sweep in Victorian London was a dirty and dangerous business. Child labour was frequently employed, as children were ideal for scurrying up and down narrow chimneys. The customary image of a chimney sweep is a young boy whose face and clothes are caked in soot, but in fact sweeps would usually work naked, as clothes would have been a hindrance, liable to get snagged or damaged on the inside of the chimney. The first law attempting to improve the working conditions for chimney sweeps was passed in 1788, and it required that each chimney sweep could have no more than six apprentices, and these apprentices had to be at least eight years old. It's astonishing to think that we're just a few generations removed from a society which considered it a mark of humane progress to only allow children aged eight and above to undertake dangerous physical labour.

Unsurprisingly, spending days trapped in narrow, soot-filled chimneys carried a range of health risks. One such condition was known as 'soot warts', which were blackened sores which would appear on the scrotum, and then spread. There was no effective treatment, and victims would often have to have their scrotum removed, but this would often result in infection and death. At the time, soot warts were thought to be a kind of sexually transmitted disease, perhaps related to syphilis. However in 1775 Percivall Pott established that the condition, which was unusually common among chimney sweeps, was a type of cancer, caused by exposure to carcinogenic soot.

This was an important discovery, not just for chimney sweeps, but for medicine in general, as this was the first time that cancer had been demonstrably linked to an external factor. Before Pott's discovery, cancer was generally thought to be a systemic disease, caused by an excess of black bile. Thanks in part to Pott's work, there were increasing efforts to regulate chimney sweeping and improve working conditions, although children continued to be used for another hundred years.

There are still many active chimney sweeps working today, although they rarely climb inside the chimney these days, and nakedness is almost certainly a thing of the past. Nowadays, chimney sweeps are more likely to describe themselves as 'chimney technicians', whose services extend beyond cleaning and unblocking chimneys, to building and repointing chimney pots, repairing fireplaces, and fitting birdcages. There is also a long-held tradition that if a chimney sweep shakes the bride's hand or blows her a kiss on her wedding day this will bring her luck, and many chimney sweeps hire themselves out for just this purpose.

Did the Great Fire of London wipe out the plague?

The conventional view used to be that it was the Great Fire of London in 1666 that finally wiped out the plague. Epidemics of plague had recurred in London about once every generation since the mid-14th century, but after the Great Fire there were no more significant outbreaks. The plague is believed to have been caused by a bacterium called Yersinia pestis, which is transmitted via fleas and rats. Thus, the theory goes, when the Great Fire wiped out London's population of fleas and rats, and destroyed the dirty, infested wooden buildings where they bred, it therefore wiped out the plague, which never returned to London.

However, there is now some doubt about this theory, for a number of reasons. The plague seems to have disappeared from most European cities at around this time, including cities which had no equivalent large-scale fires. This suggests that the plague was already somehow losing its virulence, perhaps because after each bout of plague, the survivors and their offspring were increasingly likely to be immune. The evidence does indeed suggest that survival rates tended to be slightly higher during each fresh bout of plague.

A further problem with the theory is that the Great Fire only affected the City of London, which was a small, enclosed area at the centre; the fire didn't reach the poorer suburbs. Only about a sixth of London's population lived in the City, the rest lived in the suburbs, and it was these suburbs which tended to be most affected by plague. In the Great Plague of 1665, the vast majority of deaths occurred in the suburbs, and not the City. Since the plague chiefly affected the suburbs, rather than the City, but the fire was only in the City, and not the suburbs, it's hard to see how the fire in one place could have wiped out the plague in another.

Another objection to the standard theory is that the fire was quite slow moving. It lasted for five days, and initially it was well enough contained for the Lord Mayor to dismiss it and go back to bed, commenting, 'A woman might piss it out.' His approach was clearly wrong, and later rightly vilified, but it does give an indication of the slow rate of the fire's growth. For the first few days, people didn't even try to flee the walled City. Instead, those who vacated their homes simply took a few possessions and moved to a safe area, fully expecting to be able to return soon after. Samuel Pepys even found time to bury a cheese in his garden for safe-keeping. Presumably, most of the City's rats and fleas would have been equally capable of escaping such a slow-moving fire, so it's unlikely to have killed that many of them.

There is another enduring myth about the Great Fire of London, which is that it only killed a handful of people, but this too seems dubious. It's true that only five or six people are recorded as dying in the fire. However, the deaths of the lower classes tended not to be recorded at this time. There were also accounts of foreigners being lynched by angry, rampaging mobs, as the fire was thought to be a Catholic plot, but none of these deaths are recorded. The firestorm reached extraordinary temperatures, thanks to the combination of narrow streets, over-hanging jetties, wooden buildings, and stores of gunpowder, fuel, tar, and other combustible materials. It became hot enough to melt the piles of imported steel lying along the wharves. It was hot enough to melt the iron chains and locks on the City gates. If the fire was hot enough to melt steel and iron, it was also hot enough to completely vaporise the bodies of anyone left behind or trapped, leaving no skeletons or remains.

The fire is also likely to have caused many deaths by slightly less direct means than people getting trapped and burned alive in the inferno. Many people were made homeless by the fire, and had to live in unsanitary and dangerous refugee camps. Many would have suffered from smoke inhalation, hunger, hypothermia, and burns. It's probable that significant numbers of the old, infirm, and children would have died as a result, in the weeks and months following the fire. In one of the key eye-witness accounts of the fire, John Evelyn describes the 'stench that came from some poor creatures' bodies.' It's not certain that this refers to dead bodies, but it seems likely. One estimate suggests that the fire probably caused hundreds of unrecorded deaths, and perhaps even thousands.

Why do we keep vials of smallpox?

The eradication of smallpox is regarded as one of medicine's greatest triumphs. Smallpox had blighted humanity since ancient times, consistently killing around 30–35% of those who became infected, with no treatment ever successfully developed. In the 20th century alone, it is estimated to have caused somewhere between 300 million and 500 million deaths. Even as late as 1967, around 15 million people contracted the disease, with 2 million deaths. The last epidemic of smallpox took place in Bangladesh in 1975, but was quickly contained. In 1978, the last two recorded cases occurred at the University of Birmingham in England, but were later determined to be laboratory errors. In 1979, the World Health Organisation announced that smallpox had been completely eradicated.

However, 'eradicated' may not mean quite what you think it means. It doesn't quite mean that smallpox no longer exists; it just means that there are now no living carriers – smallpox only infects humans, so there is no possibility of there being any infected animals either. There are still some confirmed vials of smallpox stored in government labs at the Centers for Disease Control and Prevention in Atlanta, Georgia, and the State Research Center of Virology and Biotechnology in Koltsovo, Russia.

Until fairly recently, there was a global consensus that these remaining vials should be destroyed, to rid the world of this potential blight. Bill Clinton signed a World Health Organisation plan to destroy the last remaining stocks in 2002. However, under George W. Bush, the US changed its stance, due to fears about smallpox being used in bio-terrorism. At the time, anti-terror experts believed that a number of volatile nations might have got their hands on quantities of smallpox, possibly including China, Iran, Israel, North Korea, Serbia, and Pakistan. There are also convincing reports that, in the relatively recent past, Russia secretly manufactured 20 tons of live smallpox virus, to be used as a biological weapon, and it's not clear what has happened to this.

The US may be more vulnerable to smallpox today than it has been for perhaps 150 years. Current stocks of smallpox vaccine are believed to be low, and they may well be deteriorating in quality, or even ineffective. The US stopped routinely vaccinating people in 1972, so few people born after that year are protected, and the immunity of many older people may also have faded, as it's not clear how long smallpox immunisation lasts. Thus, rather than destroying the remaining vials, the US plan now seems to be focused on retaining the samples, while developing a more effective vaccine, as well as developing antiviral drugs to actually treat the disease.

What is the 'disease of kings'?

The answer is gout, which has traditionally been regarded as an affliction of the wealthy and privileged, due to its association with rich foods and alcohol. Gout is an unpleasant, painful condition caused by an excess of uric acid in the blood. This acid crystallizes, forming deposits in joints, tendons, and surrounding tissues, leading to painful, tender red swellings, most often at the base of the big toe. Because of gout's association with the wealthy and well-fed, newspaper cartoons of the Industrial Age were often populated with portly, red-faced gentlemen, hobbling with painful, bandaged feet. In particular, gout has long been thought to result from regular consumption of port. Because of the prevalence of gout, the gentleman's clubs of London's Pall Mall would routinely be furnished with a number of small, velvet, tasselled footstools, so that the painfully tender toes of gout sufferers could be elevated, and thus somewhat relieved.

Bizarrely however, a recent study came to the conclusion that in fact drinking port could lower your risk of gout, as the condition was far more likely to be caused by drinking beer. Moreover, gout does not only affect the most eminent humans. Scientists have recently discovered that Tyrannosaurus Rex, the king of the dinosaurs, also suffered from gout.

CHAPTER THREE
DODGY DIAGNOSIS

'Optimistic lies have such immense therapeutic value that a doctor who cannot tell them convincingly has mistaken his profession.'

George Bernard Shaw

What was 'hepatomancy'?

Hepatomancy, which is also known as hepatoscopy, was an ancient medical system devised in Mesopotamia, in modern day Iraq. The way it worked was that patients would be diagnosed by a physician, who would make his judgements not by inspecting the patient, but by inspecting the livers of some dead animals that had been sacrificed for this purpose.

The liver was believed at this time to be the source of our blood, and therefore the source of life itself. Following this logic, people believed that the will of the gods could be divined by inspecting the livers of sacrificed sheep. You may think this sounds absurd, and you'd be correct. Each section of the liver was believed to correspond to a particular deity, and clay models of sheeps' livers, which were presumably used by doctors to help them to interpret the signs, to make their diagnosis, have been found dating back as far as 2050 BC.

Of the four 'humours', what was special about 'melancholy'?

The Hippocratic *corpus* held that human health depended on the body's four types of fluid, which were called the four 'humours'. It was believed that every aspect of our health could be explained by the balance of these humours, which were: blood, phlegm, yellow bile, and black bile - also known as 'melancholy'. Each humour was believed to play a crucial and distinct role in determining the body's health. Blood was the source of strength and vitality. Bile was the gastric juice, which was needed for digestion. Phlegm was seen as a lubricant and coolant. And black bile was responsible for darkening the other fluids, as evidenced on those occasions when blood or stools become darker.

These four humours were believed to be at the centre of a complex network of signs and relationships, which were thought to explain every aspect of a person's health and personality. For example, each humour was associated with one of the four elements that the Greeks believed comprised the universe: air, fire, water, and earth. Blood was hot and agitated, like air. Yellow bile was hot and dry, and thus connected to fire. Phlegm was linked to water, and black bile was cold and dry, like earth.

These analogies were extended to link the humours to astrology, the seasons, and different character types. Phlegm, for example, was considered to be cold and wet, which meant it was associated with winter, which is when people catch colds and chills. People with excess blood were regarded as lively, confident, and robust, or 'sanguine', while those with too much phlegm were lazy and easy-going, or 'phlegmatic'.

So what was special about melancholy? The answer is that the only one of the four humours that doesn't actually exist. Our bodies do contain blood, phlegm, and yellow bile, and these could be easily observed by doctors then and now. However, with no understanding of why blood, urine, or faeces might get darker at times, or why some people have darker complexions than others, the Greeks assumed that some as yet undiscovered fluid must be responsible, and so credited these phenomena to an unseen substance called black bile, or melancholy.

What is 'laudable pus'?

Before bacterial infection was properly understood, almost all surgical wounds would become infected. As a result, the important distinction seemed to be between those wounds which produced a creamy, yellow pus, known as 'suppuration', and those which only produced a thin, watery discharge. Since the time of the great Roman doctor Galen (129–200 AD), doctors had believed that this creamy suppuration was indispensable for healing, as pus derived from poisoned blood that needed to be expelled from the body. This view held sway until the late 1800s because, in the centuries before germ theory, it seemed to make sense. Although wounds which produced yellow pus tended to take months to heal, physicians noticed that those patients had a higher survival rate than those whose wounds were pus-free. Naturally enough, doctors drew the conclusion that yellow pus was good for you.

As a result, doctors would actively seek to encourage infections after surgery or injury. One early medical guide advised doctors to 'get wool as greasy as can be procured, dip it in very little water, add one third wine, boil to good consistency' and then insert it into the wound. Even in the late nineteenth century, it was widely accepted that cancer patients would benefit from an infection to heal the wound, after cancer surgery had taken place.

Some doctors went even further, recommending gangrene. The Parisian doctor Stanislas Tanchou argued, 'It is remarkable that...gangrene has caused the largest number of cures. Gangrene may be considered as a therapeutic agent, whether it occurs spontaneously or is induced medically.' In truth, the only thing that gangrene did regularly cause was for people's arms and legs to be amputated.

Why are doctors in old pictures often holding up a flask?

Throughout much of early history, doctors would conduct their diagnosis largely by inspecting the patient's urine. The practice is known to have been carried out as far back as ancient Egypt, Babylon, and India. Urine diagnosis, or 'uroscopy' as it is also known, was not a central plank of Hippocratic medicine, which relied more on observations made by inspecting and touching the patient, but it became a widespread practice in medieval Europe. If a doctor is present in a medieval painting, he will often be holding a large, bulbous flask up to the light, as this flask was as much the symbol of his profession then as a stethoscope and a white coat are today.

Urine diagnosis was thought to be such an effective method that patients were not even required to visit the doctor themselves; often they would simply send a urine sample from their sickbed, to be diagnosed in their absence. Doctors would do this, if possible, by inspecting the urine in various different states, firstly when it was hot and fresh, and then later when it had cooled down, and had taken on a different consistency.

Doctors used detailed colour charts against which they would compare the urine, to diagnose a wide range of conditions, including kidney disease, jaundice, and urinary tract infection. Then, on the basis of these observations, along with the patient's astrological chart, they would determine which humoural imbalance needed to be rectified. Thick urine could mean the patient was suffering from dropsy, colic, or an excess of phlegm; dark urine indicated an excess of black bile; and so on. However, not all conditions could be diagnosed by sight alone. The smell of the urine was also thought to be instructive, as was the taste. Diabetes, for example, makes the patient's urine sweet, so doctors would routinely taste their patients' urine. Doctors really did earn their money in those days.

Today we know that there is only very limited information that can be learned from a person's urine, and urine inspection is not a central plank of diagnosis. However, doctors do still analyse urine, testing its concentration and pH, as well as testing for the presence of drugs, blood, sugars, fats, bilirubin, and hormones. And in fact some of these tests can help to diagnose some of the very conditions that the medieval doctors were looking for, including kidney stones, urinary tract infections, and diabetes. Chemical analysis of urine is even undergoing something of a resurgence as a field of study, under the rather fancier name 'metabolomics'. However, tasting the patient's water has fallen somewhat out of fashion.

What was a Plague Doctor?

During the 14th century, the Black Death swept through Europe, killing around 30–50% of the continent's population, in one of the most devastating pandemics in history. There were three forms of plague: pneumonic plague, which was characterised by fever and bloody phlegm; septicemic plague, which caused fever and patches of purple skin; and the most common form, bubonic plague, which produced grotesque, pus-filled buboes to appear around the groin, neck, and armpits. All three forms were highly fatal and highly contagious. The plague returned about once every generation until the 17th century (when, as we have seen, it ended around the time of the Great Fire of London, but not because of it).

When the plague arrived in a town or city, the wealthy would flee, believing that the disease was caused by miasmas of bad air. Knowing that they could do nothing for plague victims, most doctors would also flee, and so in their place 'plague doctors' would be appointed. These were not really doctors at all; their duties generally consisted of simply visiting the homes of the sick, to assess whether or not they had the plague. If so, the house would have to be quarantined, with no one allowed in or out, and a red cross marked on the door. Although they were not qualified, plague doctors were often very well paid, to compensate them for the considerable risk they faced of becoming infected themselves.

To mitigate this risk as far as possible, plague doctors wore a protective suit which has become an iconic, sinister image of plague. The suit usually comprised a long, black, high-necked overcoat, which extended to the feet, and leather breeches, similar to a fisherman's waders, to protect the legs and groin from infection. A wide-brimmed hat was worn, as this was a symbol of doctors. Most distinctively of all, plague doctors wore red glass eyepieces, and a kind of long-nosed gas mask, in the shape of a bird's beak. The red eyepieces were meant to ward off evil, while the beak was designed to protect the wearer from the 'bad air' that was thought to cause the plague.

The end of the plague doctor's beak-like mask was often filled with herbs and spices, in the hope that their strong smell might overcome the miasmas. The whole suit would be covered with suet or wax, as a further protection. Finally, plague doctors would carry a wooden cane, which may have been used to gesture to the patient's family (as normal conversation must be fairly difficult in a wax-covered gas mask, with a beakful of herbs), or to physically move the patient's body.

Why did Claude Bernard's wife leave him?

Claude Bernard was a leading French scientist, who is regarded as a pioneer of physiology, and a key figure in the history of medicine. He discovered the role of the pancreas in digestion, and the production of glycogen by the liver. He succeeded in bringing the rigour and experimental method of science to medicine, which at the time was still largely based on superstition, anecdotes, and received wisdom. When he died in 1878, Bernard was given a public funeral, becoming the first French scientist to be granted such an honour.

Throughout his career, Bernard argued for the importance of experimentation and observation. This was controversial, particularly when it came to Bernard's approach to 'vivisection', which means experimentation on live animals (its name comes from the Latin words for 'live cutting'). Many of Bernard's contemporaries were horrified by the cruelty involved in experimenting on live animals, which was often conducted without anaesthetic. Bernard recognised that animal testing was deeply unpleasant, but he felt it offered life-saving results which were too important to ignore. He wrote, 'The science of life is a superb and dazzlingly lighted hall which may be reached only by passing through a long and ghastly kitchen.' Others, however, felt that such horrors could never be justified.

In 1845, Bernard married Françoise Marie Martin, who was known as Fanny. It was a marriage of convenience, as Fanny came from a wealthy family, which meant the match allowed Bernard to pursue his scientific career. As time progressed, Fanny became increasingly disgusted by Bernard's practice of vivisection, as did the couple's daughter. In 1869, the couple were officially separated, with Fanny citing his cruelty to animals as a major reason for the break-up of the marriage. While her former husband continued his career, Fanny went on to become a vocal campaigner against the use of vivisection.

Why was Queen Victoria's hernia never diagnosed?

In 1881, Sir James Reid was appointed as the personal physician to Queen Victoria, who was then a reasonably healthy, if somewhat overweight, 62-year-old. Victoria was a hypochondriac, so the pair had a close and involved relationship. The queen would summon Reid six times a day, and keep him intimately briefed on every detail of her constitution. On occasions, she even demanded he return early from his holidays. While he was on his honeymoon, Victoria wrote him a note to brief him on the important news that, 'The bowels are acting fully'. Even after he got married, Reid was not allowed to live with his wife, as Victoria needed him to be at her constant beck and call. When Victoria died, Reid was given a task of the highest secrecy and delicacy: he was to put a lock of hair from Victoria's close companion John Brown into her hand, along with a photograph of Brown. Victoria and James Reid clearly had an extremely close, trusting relationship.

After Victoria's death in 1901, Reid inspected her body, and found that she had a hernia, and a badly prolapsed uterus, both of which had never been diagnosed. The reason they had never been diagnosed was that Reid had never been allowed to see Victoria with her clothes off, as royal propriety forbade it. He had never been allowed to place a stethoscope on her chest, and he never once saw her in her bed until six days before her death.

This kind of primness sounds like the kind of thing that could only come about through the absurdities of royal protocol, but in fact it was fairly typical of how normal doctors of the period operated, and had done for centuries. Galen had recommended taking the patient's pulse at the wrist, specifically because this did not require any undressing. Through the Middle Ages, doctors seemed to gradually retreat from physical contact with their patients. The 18th century German doctor Johannes Storch, who was a leading medical authority and widely published, would rarely even meet his patients. Instead, he carried out most of his diagnoses and treatment through correspondence and intermediaries. As late as the 1890s, some family doctors in the US would do nothing more intimate than taking the patient's pulse and inspecting their tongue.

What's the difference between a moron and an imbecile?

The short answer is: about 25 IQ points, or at least that used to be the answer. Today, the words 'moron' and 'imbecile' mean essentially the same thing: they are both pejorative, insulting terms, which simply mean 'person of low intelligence'. However, this was not always the case. In the first American IQ tests, 'moron', 'imbecile', and 'idiot' were neutral, inoffensive terms, which each had a specific and distinct meaning. A moron was an adult with an IQ of between 51 and 70; an imbecile was somewhat slower, with an IQ between 26 and 50; and an idiot was the slowest of all, with an IQ between 0 and 25.

The very first IQ test was designed for a specific, practical purpose. It was developed in France by two psychologists, Alfred Binet and Theodore Simon, and thus became known as the Binet-Simon scale. The system comprised a series of thirty tasks of increasing complexity. The simplest tasks included asking the child to shake the examiner's hand, or to follow a lit match with their eyes. The most difficult tests included remembering a series of seven random numbers, and finding three rhymes for the word *obeisance*, which is presumably easier if you are actually French, as these children luckily were. The tests were based on observations of an average child's level of ability at each age, and thus created the concept of a 'mental age', as distinct from the child's actual age.

As I mentioned, the purpose of this system was a purely practical one: to provide a fair and consistent way of measuring which schoolchildren had special learning needs, so that these children could be educated separately. The invention of the Binet-Simon scale was thus designed to address an apparent problem in schools, as some teachers were suspected of choosing to exclude children who were troublesome, but not in fact developmentally challenged.

Binet was careful to recognise the limitations of his system, as he believed that intelligence was not something inherent and fixed, but much more fluid and complex. He wrote:

> 'I do not believe that one may measure one of the intellectual aptitudes in the sense that one measures length or a capacity. Thus, when a person studied can retain seven figures after a single audition, one can class him, from the point of his memory for figures, after the individual who retains eight figures under the same conditions, and before those who retain six. It is a classification, not a measurement. We do not measure, we classify.'

However, when the system crossed the Atlantic, it was taken up by groups with very different goals. The first translation of Binet's system was carried out by Henry H. Goddard, who was a leading psychologist. He was also a eugenicist, who believed that society should be organised and controlled by the most intelligent people, and that those of the lowest intelligence should be institutionalised or sterilised. Goddard developed his own IQ tests, based on Binet's scale, which categorised adult participants as gifted (IQ of 130+), normal (71–129), moron (51–70), imbecile (26–50), or idiot (0–25).

Now, while there are many things we might criticise Goddard for, his use of language shouldn't be one of them. At the time, the words 'moron' and 'imbecile' were not insulting terms. In fact, the word 'moron' didn't even exist. Goddard invented it in 1912, from the Greek word 'moros' which means 'dull'. It was only because these tests became common and widely known that these words took on a pejorative meaning in the wider language.

This shift in meaning illustrates an interesting concept called the 'euphemism treadmill', in which words which are initially inoffensive, but which refer to something which is perceived negatively, will gradually begin to acquire a derogatory, taboo meaning. In 1912, 'moron' was not an insulting word, but by the 1960s its meaning had changed, and so psychologists began using new, politically correct terminology, describing these lower categories with the more enlightened language of mild, moderate, severe, and profound retardation. That's right, 'retardation'. As you've no doubt already noted, in time that word also became unsuitable, as did the next politically correct idiom: 'mentally handicapped'.

Today, the proper form is to refer to a person as having 'special educational needs' or 'learning difficulties', constructions which are in line with our current view that people should not be defined by their condition, and that we shouldn't automatically regard difference as a handicap, so instead we now refer to people as having 'needs or 'difficulties', rather than being the thing in question. However, it seems that no matter how careful we are to try and design neutral, inoffensive language for sensitive issues such as this, the meaning of this language is always likely to shift and become pejorative in time.

What is a lancet?

A lancet is a medical device which is used for bloodletting. From the time of Hippocrates until the 19th century, illness was generally believed to be caused by an imbalance of the four humours, and as a result the main tools at a doctor's disposal remained roughly unchanged. These tools included 'purging' - which means induced vomiting, sweating, enemas, and perhaps most commonly of all, bloodletting. Bloodletting was so central to medicine for so much of history that the name given to the leading medical journal was 'The Lancet', and of course that publication still carries the name today.

There were various types of lancet, including spring-loaded lancets, thumb lancets, fleams, and scarificators, but they all essentially consisted of a number of small blades or spikes, which were used to cut into a vein or artery. As well as lancets, there were various other devices used for bloodletting. Some doctors used 'fire cupping', in which a small bowl would be heated, and then held against the skin at the site of incision, creating a vacuum that would draw out the blood. Many doctors used leeches, to such an enormous extent that France had to import millions of the creatures, having harvested its own leech population to the brink of extinction.

Bloodletting was recommended by doctors for almost any ailment. Blood was believed to originate in the liver, and to then grow stagnant as it built up in the body. Consequently, blood loss was generally seen as a good thing, and nosebleeds, varicose veins, and haemorrhoids were all encouraged, because they helped to get rid of excess blood. Patients were often bled until they fainted. If the patient hadn't fainted, that was usually a sign that you hadn't taken enough blood.

Some of the uses of bloodletting were particularly absurd. Galen's recommended cure for a patient suffering from blood loss was, you guessed it, bloodletting. Before any surgerical procedure, patients would have blood removed, as would mothers before childbirth. Before an amputation, doctors would estimate the amount of blood that the limb contained, and then remove it, before then removing the limb (which was, of course, still full of blood).

In the vast majority of cases, bloodletting did considerably more harm than good, and many patients died as a result. What's astonishing is that bloodletting continued to be used and advocated, even as the science of medicine advanced, and the theory of the humours increasingly became redundant. In 1628, William Harvey demonstrated that blood circulated around the body, effectively disproving the logic of bloodletting, yet the practice continued. In the 1830s, bloodletting was shown to be useless for pneumonia and fevers, for which it was commonly employed, but despite this evidence doctors persisted. In 1857, John Hughes Bennett produced detailed statistical evidence comparing bloodletting with both non-treatment and placebo treatment, and clearly demonstrated that non-treatment was better than bloodletting. Even then, the practice of bloodletting endured.

With each advance in medical knowledge, bloodletting became increasingly anachronistic and yet it persisted, because although new knowledge was being rapidly discovered, there were hardly any new cures or treatments being developed at this time, and doctors had to continue to provide some kind of service. One medical historian has estimated that, despite all the advances of the 18th and 19th centuries, there was no meaningful advance in the effectiveness of treatments until Joseph Lister's discovery of antisepsis in 1867. This may explain why, as late as 1923, a leading medical textbook was still recommending bloodletting.

However, although it was mostly harmful, bloodletting was beneficial for a handful of conditions. Edema, which used to be known as 'dropsy', is a type of fluid retention that can cause severe heart and lung problems. Today, edema is treated using diuretics and vasodilators, but in the past bloodletting did have a useful effect. Polycythemia is a disease in which your body produces too many red blood cells, and so bloodletting is still used for this. Other conditions which still call for bloodletting include the enzyme disorder porphyria, and hemochromatosis, in which your body absorbs too much iron.

Does chocolate give you spots?

There is an enduring myth that chocolate causes spots, but the evidence does not support it. A recent meta-analysis, which combined the results of 21 separate observational studies and 6 clinical trials found no link between chocolate consumption and acne. There was also no link between acne and greasy foods, such as burgers, fries, and pizza (although of course none of this undermines the general health benefits of a nutritious, balanced diet). The study did however find a possible link between cow's milk and acne, as well as foods with a high glycemic index, such as highly refined bread and cereals. Chocolate itself, however, has a low glycemic index.

Overall, the evidence continues to suggest that the main cause of acne is not diet, but hormones. During adolescence, our bodies produce more of the hormone androgen, and one of its many effects is to stimulate our sebaceous glands to produce more sebum and grow in size. Spots occur when our hair follicles get blocked with dead skin, causing sebum to build up behind the plug. 96% of teenagers are affected by acne at some point, and in many sufferers it continues into adulthood. The other important factor is simply genetics. Acne will tend to run in families, as you are likely to have a similar skin type to your parents.

Did President Harrison die of a cold because he refused to wear a hat during his inauguration?

William Henry Harrison was a notable US president for a number of reasons. He was elected in 1841, but died just 32 days later, making him the first president to die in office. His tenure remains the shortest on record by some distance, as the next shortest presidency was that of James A. Garfield, who was shot by an assassin after a full 199 days in office. Harrison is also notable for being elected with the highest share of the popular vote of any president, although this is partly down to the high voter turnout at that particular election.

At the time, Harrison was also the oldest man to become president, at 68 years old, and his detractors mocked him for this, referring to him as 'Granny'. Perhaps to emphasise his health and vigour – he was, after all, a war hero – Harrison delivered the longest inauguration speech on record, at 1 hour and 45 minutes, on March 4, 1841. It was a bitterly cold day, and it may even have been snowing, but despite the weather, Harrison refused to wear a hat or overcoat, and the entire address was conducted outside. By March 26, Harrison had fallen ill with a cold, and on April 4, he died of pneumonia.

It has often been supposed that Harrison's death was caused by his decision to deliver such a lengthy inauguration speech in the freezing, outdoor conditions, but this assumption is surely false. Harrison fell ill with his cold on March 26, more than three weeks after his inauguration, which suggests that it couldn't have been caught on that day. Cold symptoms usually begin within 2–5 days of infection.

The cold also couldn't have been caused by the weather because, contrary to the received wisdom, colds aren't caused by low temperatures, they are caused by viruses – there are more than 200 different cold viruses. Although it is true that people get more colds in the winter, that's not because we're outside in the cold. On the contrary, we catch more colds in winter because we're huddled up indoors, with the heating on and the windows shut. In the winter, we find ourselves in closer proximity to other people, who might infect us with colds, and our warm, unventilated homes are ideal conditions for viruses to thrive. This won't be of much comfort to President Harrison, but at least his death wasn't his own fault for not wearing a hat.

CHAPTER FOUR
CURIOUS CURES

'He's the best physician that knows the worthlessness of most medicines.'

Benjamin Franklin

Why do boxers drink their own urine?

In 2009, boxer Juan Manuel Márquez allowed HBO's cameras to follow him as he trained for his fight with pound-for-pound champion Floyd Mayweather Jr. At one point in the show, viewers saw Márquez drink a glass of his own urine, which he explained was a regular part of his training regime. According to the boxer, urine contains lots of vitamins, which are expelled by the body, and so he drinks it for his health.

In fact, drinking urine, or 'urine therapy' to give it its proper name, is surprisingly common among fighters. The Brazilian Mixed Martial Arts light-heavyweight Lyoto Machida drinks his own urine every morning, as do his whole family. Fellow MMA fighter Luke Cummo also advocates urine therapy, claiming it gives him the edge over his opponents, although since his record so far is six wins and six defeats, he may not be the best ambassador for the practice.

After winning his WBC world heavyweight title fight against Nigerian Samuel Peter in 2008, Vitali Klitschko revealed his own special use for urine. Klitschko wraps his baby's wet nappies around his fists after a fight, to prevent swelling. He explained, 'Baby wee is good because it's pure, doesn't contain toxins, and doesn't smell,' At the time of writing, Klitschko is still the WBC champion, so perhaps it does help.

In fact, urine drinking is not just practised by punchdrunk pugilists. Many devotees of alternative medicine swear by the practice, which has gone on for thousands of years. Indian holy men would drink their own water, as well as massaging it into their skin. The Koran bizarrely recommends the drinking of camels' urine, for its medicinal properties. In China, the urine of young boys is drunk by adults, as it is believed to have health-giving properties, and babies' faces are washed in urine to protect their skin. And the practice continues today. As recently as 1978, the Prime Minister of India was publicly advocating urine therapy as the ideal solution for millions of poor Indians who could not afford to see a doctor.

So, is there anything to it? Sadly (or not, I suppose, depending on how keen you are to drink urine) the answer seems to be no. The practice is fairly harmless, as urine is generally sterile, and any infection that might be carried in the urine would already be present in the patient's body, assuming they only drank their own urine rather than somebody else's. Márquez is right to say that urine may contain vitamins, but these are vitamins that the body has already expelled, because it has no use for them. The only meaningful effect of drinking urine is likely to be a bowel movement, or even diarrhoea, as the urea in urine has a laxative effect. Overall then, there is no evidence that any form of urine therapy is beneficial. And as for the fight, Márquez lost on a unanimous points decision, 120–107, 119–108, 118–109, which means that Mayweather won almost every round.

But what about jellyfish stings? Everyone knows urine is good for them ...

Well, you've probably seen that episode of 'Friends' in which Joey and Chandler have to pee on Monica's leg, as this is known to be the best way to ease the pain of a jellyfish sting. However, despite this advice being so well known, in fact the evidence seems to suggest that urine doesn't help to treat a jellyfish sting at all, and it might even have made Monica's pain worse.

A jellyfish's tentacles are covered with tiny stinging cells called nematocysts. The jellyfish fires these into its prey at enormous speed, injecting a powerful venom. If you get stung by a jellyfish, there are two main priorities. Firstly, any remaining parts of the tentacle have to be removed, without triggering the nematocysts to release more venom. This is tricky, because nematocysts will release venom in response to fresh water, certain chemicals, or physical pressure, so one recommendation is to use a towel or sand to carefully remove any remaining parts of the tentacle. Touching the tentacle with your bare hands is not a good idea, as it will only result in more stings.

The second job is to wash away the remaining nematocysts, but this is a difficult task. Fresh water will make them release more venom, as they react to a change in the concentration of salts. Various other liquids have been recommended, including Coke, red wine, and ammonia, but there is very little evidence to support any of them. As you might imagine, people are not exactly queuing up to take part in scientific tests involving getting stung by a jellyfish (although, amazingly, some volunteers have been found). Based on what evidence there is, it seems that vinegar or its key ingredient, weak acetic acid, are perhaps the best choice for washing the sting, as this acid is known to neutralise the nematocysts of many types of jellyfish, including the deadly box jellyfish. However, some other types of jellyfish react badly to vinegar, making the sting more painful, so seawater may be preferable in those cases.

The theory that urine helps to ease the pain of a jellyfish sting may have developed because urine is usually mildly acidic. However, the acidity of urine is quite weak, and it's likely therefore that the nematocysts would react to urine as if it were fresh water, meaning that it would make them release more venom. Therefore, washing a jellyfish sting with urine is not recommended.

Once the skin has been rinsed, either with vinegar or seawater, the remaining nematocysts can be removed by shaving the area, using a razor or a knife. Most jellyfish stings are painful, but not fatal, and the pain usually disappears after about 24 hours. The shame of having your best friends pee on you, on the other hand, is likely to sting for much longer.

Why was urine used to wash battlefield wounds?

Urine has also been used for centuries as an antiseptic, to clean battlefield wounds, when no clean water or other antiseptics were available. It may sound unpleasant, but urine is usually sterile, unless the donor has a urinary tract infection, which means it was probably a reasonably effective solution, so to speak.

Urine was certainly preferable to some of the alternative battlefield balms. Before the 16th century, doctors would often pour boiling oil onto serious battlefield injuries, to cauterise the wound and stop the bleeding. One day, the great French surgeon Ambroise Paré was tending the wounded on a battlefield, when he ran out of boiling oil. He remembered an old ointment he had read about, which had supposedly been used in Islamic medicine, and decided to give it a try. He treated the remaining patients with this ointment, and then left for the night.

When he returned the next day, Paré found that he had inadvertently performed a scientific experiment. The soldiers who had been treated with boiling oil were still in agony, and showed no improvement in their condition. However those who had received the soothing ointment were much improved, thanks largely to the antiseptic properties of the turpentine it contained.

Was Adolf Hitler addicted to crystal meth?

Bizarrely, it seems that he was. Hitler was a hypochondriac, and every day he would have his doctor Theodor Morell give him vitamin injections into his buttocks. As World War Two progressed, Hitler was having as many as five injections a day, and Morell was lacing them with methamphetamine, also known as crystal meth, to keep the Fuhrer alert and energetic. Thanks to these injections, Hitler was 'fresh, alert, active, and immediately ready for the day … cheerful, talkative, physically active and tending to stay awake long hours into the night.' Albert Speer felt that Hitler's methamphetamine addiction was one reason for his rigid tactics in the later stages of the war, when he would refuse to allow his generals to make any tactical retreats, regardless of the circumstances.

Although Hitler's drug use may sound surprising, in fact amphetamines were widely used during World War Two. The German army were fuelled by an amphetamine called pervitin, with the army supplying millions of tablets to the troops, particularly during the blitzkrieg attacks on Poland and France. The British and Japanese armies also distributed huge volumes of amphetamines to their soldiers. The US armed forces didn't officially authorize amphetamine use on a regular basis until the Korean War, but many troops stationed in Britain became familiar with benzedrine – one estimate suggests that 80 million benzedrine pills were supplied to US servicemen by the British, with a similar number being provided by US medics.

Although Hitler's drug use is well documented, there are many other theories about his health which are more speculative.

Various historians have argued that Hitler suffered from syphilis, Parkinson's disease, Asperger syndrome, irritable bowel syndrome, irregular heartbeat, skin lesions, and a range of mental health disorders including borderline personality disorder and schizophrenia. One enduring rumour held that Hitler had only one testicle, perhaps as a result of an injury he suffered during World War One, when he was shot in the groin. After Hitler's death, a Russian autopsy confirmed that he was missing his left testicle, which seemed to prove that the popular wartime song 'Hitler Has Only Got One Ball' was, amazingly, correct.

However, there are strong suspicions that the Russian autopsy was a work of propaganda. For one thing, Hitler's body is believed to have been cremated in the bunker where he shot himself, which would presumably not leave much of a scrotum to be examined. Furthermore, there is evidence that the earliest popular versions of the 'one ball' song were about Hermann Göring, with the song beginning, 'Göring has only got one ball, Hitler's are so very small ...' and so on. Although the order of the protagonists was later reversed, it would seem to be a bizarre coincidence for a song to appear, making such a bizarre and specific claim about Göring, which then turned out to be true of Hitler instead. It seems likely therefore that this particular story is a myth.

How did Agatha Christie's 'The Pale Horse' save lives?

On its release in 1961, Agatha Christie's murder mystery 'The Pale Horse' marked something of a departure from her usual, classic formula. The book didn't feature either of Christie's two celebrated detectives, Hercule Poirot and Miss Marple, and it was set in a distinctly contemporary, modern setting, rather than the standard country house which had become such a cliché of the genre. The book even features a fistfight in a pub.

Like many of Christie's novels, the book features murder by poisoning. Christie herself had worked as a nurse and a pharmacist in her youth, which may have sparked her endur-ing fascination with poisons. In the case of 'The Pale Horse', the poison of choice was thallium, which is a fairly common metal. Thallium is a very effective poison, as many of its salts are odour-less, tasteless, and easily dissolved. The effects of thallium poi-soning take time, which is also useful for a would-be poisoner, and they are difficult to distinguish from other conditions such as epilepsy, encephalitis, and neuritis. Symptoms of thallium poisoning include lethargy, numbness, blackouts, slurred speech, and hair loss – in fact, in the 1930s thallium was sold over the counter as a depilatory product, under the exotic-sounding name 'Koremlou Creme'.

In 1975, 14 years after the book's release, Agatha Christie received a letter from a woman in Latin America, thanking her for saving her friend's life. In her letter, the woman explained that she had noticed the symptoms of thallium poisoning in a friend, thanks to having read 'The Pale Horse', and had conse-quently saved the life of this unwitting friend, who was being slowly murdered by her husband.

In fact, 'The Pale Horse' seems to have saved a number of lives since its publication. In 1977, a nurse at Hammersmith Hospital in London was treating a child from Qatar who was suffering from a mysterious illness. Luckily, the nurse had also read 'The Pale Horse', and so spotted the symptoms of thallium poisoning, saving the child's life. In 1971, the notorious poisoner Graham Frederick Young was caught, thanks to a doctor connected with the case recognising the symptoms of thallium poisoning, again thanks to 'The Pale Horse'.

Why did ancient Egyptians apply mouldy bread to wounds?

From the time of the ancient Egyptians onwards, the prin-ciple that mouldy bread could be used to keep cuts clean was a common and widespread tenet of European folk medicine. Today, we know why this may have worked: it's because fungi can have antibacterial properties. This principle was discovered in 1877 by Louis Pasteur, the father of bacteriology.

Before Pasteur, scientists already knew that bacteria could be killed using chemicals, but Pasteur discovered the possibility of using biological agents, when he observed that common moulds slowed the growth of Bacillus anthrasis, the bacterium which causes anthrax. There is even some speculation that Pasteur identified this strain of mould as penicillin, many decades before penicillin is generally held to have been discovered in 1928.

What was the 'Powder of Sympathy'?

Most of what was regarded as conventional medicine in the seventeenth century would strike us as wacky and bizarre today, but even in this context of superstitious quackery, Sir Kenelm Digby's 'Powder Of Sympathy' seems wonderfully insane. The powder was designed for a very specific purpose, to heal rapier wounds (a rapier is a type of sword), and it contained a range of bizarre ingredients, including earthworms, pigs' brains, iron oxide, and powdered mummy. This last ingredient was exactly what it sounds like: bits of mummified corpses, ground into a powder. Use of powdered mummy in medicine was actually fairly common at this time. Doctors had been prescribing powdered mummy for centuries for all kinds of ailments – a practice which seems to have started after people in the 12th century observed that mummified corpses were extremely well preserved.

As if this story weren't batty enough, this salve was not even meant to be applied to the wound itself, but rather to the weapon which had caused the wound. Digby believed that applying this strange concoction to the offending sword would somehow encourage the wound itself to heal, through 'sympathetic magic'. Furthermore, Digby was not some obscure crank. On the contrary, he was a renowned courtier, a diplomat, and a leading intellectual of the time. His book on the Powder of Sympathy went through 29 editions. His other achievements included being the first person to observe that plants require oxygen (or 'vital air', as he termed it), and inventing the modern wine bottle. Digby's correspondence with the legendary mathematician Fermat contains the only surviving copy of any of Fermat's mathematical proofs – a demonstration that the area of a Pythagorean triangle cannot be a square, through Fermat's method of infinite descent.

Another use that Digby proposed for his Powder of Sympathy was that it should be used by sailors to determine their longitude at sea, which was an extremely important and as yet unsolved problem. Digby's suggestion was that a bandage should be taken from a wounded dog, which would then be taken to sea (the dog, not the bandage). At a set, pre-agreed time each day, the bandage (on land) would be placed into the Powder of Sympathy, and through sympathetic magic, this would cause the dog (at sea) to yelp. Thus, the crew on board the ship would effectively get a daily alarm call notifying them of the correct time, which would allow them to calculate the ship's longitude. It is not known whether or not this method was ever tried. Of course, one of the downsides of this method would be that, on long journeys, the dog would have be frequently reinjured, to make sure the wound remained effective.

Which disease was thought to be cured by leading the patient three times around a pigsty, while wearing a donkey's halter?

According to a book of Irish folk traditions, it was believed that an effective cure for mumps was to put a donkey's halter over the patient's head, and lead them three times around a pigsty. However, why anyone believed this might actually work remains a mystery.

What was Vin Mariani?

Vin Mariani was a health tonic, created in 1863 by chemist Angelo Mariani. Mariani had been inspired by an article which outlined the powerful effects of cocaine, a drug which had only recently been synthesised from the coca plant, and so he decided to produce a wine that would harness these effects, by lacing regular Bordeaux red wine with coca leaves. Mariani's tonic contained 6 milligrams of cocaine per fluid ounce of wine, although he later upped the dose to compete with stronger rivals.

The tonic was an enormous success, perhaps unsurprisingly. Mariani's advertising played up the tonic's health-giving properties, claiming that the drink was endorsed by 8,000 doctors, and ideal for 'overworked men, delicate women, and sickly children'. It was enjoyed by many prominent people, including Thomas Edison, Queen Victoria, the Czar and Czarina of Russia, Pope Saint Pius X, and Pope Leo XIII, who even appeared in an advertisement for the wine, and awarded it a Vatican gold medal.

Vin Mariani, an invigorating tonic wine, thanks to its special ingredient: cocaine.

Interestingly, Vin Mariani's popularity was not just down to the presence of alcohol and cocaine, two pleasurable drugs in their own right. It also benefited from the particular way in which they react with one another. When taken with alcohol, cocaine produces a potent psychoactive metabolite called cocaethylene, which is only produced by the two drugs taken together, not by cocaine on its own. Today, cocaethylene is becoming an increasing concern among health professionals, as there is growing evidence that heavy drinking combined with cocaine use is causing serious heart problems among the under-40s, because of the effect of cocaethylene.

Vin Mariani was successfully exported to the US, where it sold extremely well, and inspired a certain John S. Pemberton to come up with a similar product of his own. In 1885, Pemberton began selling Pemberton's French Wine Coca, but prohibition laws were soon passed, although of course these only restricted the sale of alcohol, not cocaine. Pemberton responded by developing a non-alcoholic version of his drink, which he decided to call Coca-Cola, and this too went on to become reasonably popular, I believe.

What is a clyster?

'Clyster' is an old-fashioned word for an enema, which is the introduction of water or some other liquid into the rectum and colon. Clysters are thought to have been carried out since the time of ancient Egypt, supposedly inspired by the ibis bird, which uses its beak to flush salt water into its cloaca. Clysters became an extremely fashionable treatment between the 17th and 19th centuries among the wealthy. They were used to treat constipation, along with a wide variety of other ailments, although they were almost certainly useless for any complaint other than constipation.

A conventional enema would consist of warm water, possibly with added salt, baking soda, or soap. Some doctors would perform more elaborate enemas, adding coffee, bran, herbs, honey, or camomile. To perform the enema, the doctor would insert a large syringe deep into the anus, beyond the sphincter muscles. The syringe would have a large bulb attached to the top (the top being the end outside the body), which would contain the enema liquid. In earlier times, a pig's bladder was often used for this purpose. The bulb was then squeezed, forcing the mixture through the syringe and up into the colon.

In high society, enemas became enormously popular, with aristocratic hypochondriacs taking three or four delicately scented enemas a day. The patient would kneel, with his or her bum in the air, to give the doctor maximum access. Some of the more refined women would try to maintain their privacy by kneeling behind a screen, blocking the doctor's view of everything but their anus, in a rather bizarre attempt to preserve their modesty.

Louis XIV of France was said to have had more than 2,000 enemas over the course of his reign. He also set up a special team of detectives to investigates the spate of murders by enema poisoning that were becoming a growing problem. On one occasion, the visiting Duchess of Burgundy shocked the King by taking an enema in his presence, as if it was the most natural thing in the world. It's little wonder that scenes featuring enemas began to appear in the comic plays of satirists such as Molière.

Enemas have also been used for other purposes. In the Middle Ages, feeble patients were given liquid nutrient enemas, to feed them if they were too weak to eat. Enemas are also sometimes used to administer drugs. If a drug is likely to upset the stomach, for example, doctors may instead use an enema. In the 18th century, there was a brief trend for tobacco smoke enemas, which were thought to be a good way to revive victims of drowning. The practice declined after 1811 when it became clear that the tobacco smoke was toxic. This exposed the fraudulent hucksters who'd promoted and sold the treatment, leading to a new idiom: 'Don't blow smoke up my ass.' Although 'clyster' is now an outdated term, enemas are still quite fashionable in certain circles, although today they are known by the trendy-sounding name 'colonic irrigation'.

What is a hemiglossectomy?

This was the name for an astonishingly brutal 18th and 19th century cure for stuttering, which involved cutting off half the tongue. Hemiglossectomies are still sometimes performed today, in cases of oral cancer, when all other treatments have failed, leaving removal of part of the tongue as the only viable option.

However today these operations are at least performed under general anaesthetic. This wasn't the case in the past.

What makes the historical use of this operation so remarkable is that it seems so absurdly excessive when used to treat something as mild as a stammer, which can be dealt with in many other ways. According to contemporary accounts, the hemiglossectomy didn't even work, as many patients who survived the operation continued to stammer, while others bled to death. In 1937, one correspondent to the British Medical Journal summed up the objections to the treatment:

> *'There is no logical reason why it should not correct the stammering habit, but the adoption of such a course indicates a complete ignorance of speech training. The hangman can cure indigestion, but there are other and better methods!'*

What causes warts? And is there a cure?

Warts are caused by the human papilloma virus, which has more than a hundred different strains. Strains 1, 2, and 3 cause the most common types of wart, while there are a number of other strains that cause genital warts. Warts are mildly contagious, and can be passed from person to person, as well as via indirect contact through towels, shoes, wet floors, and so on. Warts are often found on the hands and fingers, but they can grow all over the body, including on the eyelids, rectum, or the inside of the mouth.

Warts will sometimes heal themselves without being treated, but it's generally sensible to get them seen to. There are a number of treatments, the most effective of which is to apply salicylic acid. Warts can also be frozen with liquid nitrogen, cauterized, or removed with a laser. However, none of these treatments is thought to work more than about 75% of the time, and warts will sometimes regrow after being removed.

Warts have given rise to a great number of folk cures and theories over the years, no doubt because they seem to appear and then disappear for no obvious reason. The length of time a wart lasts is also very unpredictable, compared with many illnesses, which is likely to have encouraged people to believe that their chosen remedy has worked, if it happens to roughly coincide with the wart disappearing. By contrast, we all know that colds tend to last a few days, so when we do recover from a cold on day three, we are probably less likely to attribute our recovery to having touched a worm or buried a pebble.

Warts are of course traditionally associated with toads, because toads have a warty appearance (although the bumps on a toad's back are not actually warts). There used to be a common theory that you caught warts by touching a toad. To cure the wart, you were advised to wear a live toad in a bag around your neck, until the poor creature died. Alternatively, you might be told to rub a frog or snail on the wart, and then impale it on a twig (the frog or snail, that is, not the wart). For those looking for an even bloodier form of animal cruelty, you could try chopping off the head of an eel, dripping the blood onto the wart, and then burying the head, as this was another widely recommended 'cure'.

There were also some rather less gruesome folk remedies out there. One theory held that a wart should be rubbed with a piece of elder wood, which was then thrown away. If there was no elder wood to hand, a bean pod would do just as well. You could also sell or give away your wart. To pass it on to the dead, you would have to throw a stone after a funeral procession (a risky option, surely), or alternatively rub the wart while a funeral procession passed, or collect some mud from the mourners' shoes, and apply it to the wart. Warts could also be sold, for a nominal price, or taken to a crossroads and symbolically buried. Alternatively, a sufferer could try making a face in the mirror at midnight, or blowing on the wart nine times during a full moon. In short, pretty much any old nonsense would do, which does at least make warts an instructive example of how people will come to believe in almost any remedy, as long as it is occasionally seen to roughly coincide with the odd person getting better.

Why has arsenic been so widely used in medicine?

Arsenic is one of the world's most notorious and widely used poisons. Nero secured control of the Roman Empire by poisoning his stepbrother Britannicus with arsenic. In 15th century Italy, the Borgias used arsenic to murder their political opponents. In 17th century France, its usage was so common that it became known as *poudre de succession*, which means 'inheritance powder'.

Arsenic is an ideal poison because it is odourless, tasteless, and easily assimilated into food and drink. Until the 19th century, there was no test for its presence in the body, and because its symptoms resemble those of cholera, the resulting deaths would be unlikely to raise suspicion. Suspected victims of arsenic poisoning include Napoleon Bonaparte, King George III of England, and Francesco I de' Medici, Grand Duke of Tuscany.

The effects of arsenic poisoning are varied and unpleasant. One clue is that victims of arsenic poisoning will often have a garlicky smell to their breath or sweat. Mild symptoms may include a skin rash, hair loss, and white lines on the fingernails. In more severe cases, victims will suffer bloody vomiting, abdominal pain, severe diarrhoea, and dehydration. One form of arsenic poisoning will even turn the patient's urine black.

However, arsenic has also been used for centuries as a medicine. In traditional Chinese medicine, it is known as *Pi Shuang*, and still used to treat cancer. Arsenic was a key ingredient in numerous patent medicines, including Fowler's Solution, a very popular tonic which was prescribed from the late 18th century until as recently as the late 1950s, as a cure for malaria and syphilis. Donovan's Solution, containing arsenic, was used to treat rheumatism, arthritis, and diabetes. Victorian women also used it as a cosmetic on their arms and face, to improve their complexion.

Since arsenic was well known to be poisonous, why was it used so often as a medicine? How could anyone think that arsenic would be good for them? One reason may be that mild arsenic poisoning breaks down the blood vessels in the face, thus appearing to give the patient a healthy, reinvigorated glow. Secondly, the idea of a poison also being a medicine is not as daft as it might first sound. We naturally tend to think of poisons and medicine as being two distinct, opposing concepts, but in fact they are often one and the same. Many medicines work by attacking or poisoning harmful agents in the body, such as parasites and bacteria. In many cases, these medicines are also somewhat toxic to us too, but usually the benefits outweigh the risks.

With this in mind, it is not entirely surprising that arsenic has been used in many medicines which have actually proven to be effective, some of which are even still in use today. After its discovery in 1909, Salvarsan became the most widely prescribed drug in the world, as at the time it was the most effective available treatment for syphilis (although it became supplanted in the 1940s by penicillin). Arsenic is still used today in chemotherapy, to treat certain forms of leukaemia. It is also believed that arsenic may prove to be effective in treating autoimmune diseases.

CHAPTER FIVE
THE GOOD DOCTOR

'My doctor gave me two weeks to live. I hope they're in August.'

Rita Rudner

How did medical student Vesalius get his big break?

In 1536, the 22-year-old Andreas Vesalius was a lowly medical student at the University of Louvain in his native Belgium. However, he had just returned from studying at the University of Paris, and found the standard of teaching at Louvain dismally inadequate. On one occasion, seeing his lecturer's clumsy attempt at dissecting an animal, Vesalius strode to the front of the class and showed him how it ought to be done.

What may have been the defining event of Vesalius's life took place soon after. Outside the city walls, he found the body of an executed criminal swinging from a gallows. The skeleton was intact, and held together by ligaments. Excitedly, he tore off the arms and legs and rushed home with them. He returned later that night, breaking the city curfew, to climb the gibbet, smash the chain, and remove the rest of the body. He then boiled up the bones, and used them to create his first skeleton. When anyone asked, he lied and said that he had brought the bones with him from Paris.

At this time in Northern Europe, the idea of dissecting a human body was so shocking as to be practically unthinkable, which meant the science of anatomy could make only very limited progress, as the subject was dominated by dubious received wisdom, which could never be tested and questioned. An anatomy lecture at the time consisted of studying the ancient texts of Aristotle and Galen, perhaps illustrated by a dissection of an animal such as a dog or pig, on the assumption that all animals' bodies and organs were basically the same. The focus of the lecture was the book, which was the true source of knowledge; the dissection was merely a sideshow.

Vesalius's revolutionary approach was to question the hallowed teachings of Galen, and conduct his own observations. In doing so, he found that the great Galen had actually made numerous errors. Galen believed the human jaw consisted of two bones, but Vesalius found it was just one. Galen believed the breastbone contained seven bones, but Vesalius found just three. Galen contended that women had one more rib than men, but Vesalius couldn't find it. In total, Vesalius detailed more than 300 such errors, and came to the conclusion that Galen had never actually dissected a human body; instead Galen must have simply dissected other animals, such as dogs, pigs, and monkeys, and assumed that human bodies were the same (it seems likely that this is indeed what Galen had done).

Vesalius's anatomy lectures became a sensation, drawing large crowds of spectators. As interest in his work grew, he was granted the corpses of executed criminals to dissect. Often, he would dissect a human corpse side by side with a live animal, which he would vivisect simultaneously. The idea was to demonstrate the action and function of each part of the body as far as possible in the live animal, while identifying the equivalent part of the human corpse, and comparing their similarities and differences. Vesalius seems to have taken a rather gruesome pleasure from vivisection. For example, he described his satisfaction at the vivisection of a pregnant bitch, as he could clearly see the unborn puppies struggling to breathe, as the placental blood supply was cut off.

Today, Vesalius is regarded as the father of anatomy, and one of the most important figures in the history of medicine. His illustrated guide to human anatomy, 'De Humani Corporis Fabrica', was enormously influential, and he made a vast number of important discoveries, particularly concerning the structure and functions of the heart and circulatory system. However, just as important as these insights was Vesalius's use of the scientific method, in which experimental evidence is given more weight than the authority of the great figures of the past.

Vesalius's pratices generated considerable controversy throughout his career. At the age of thirty, he was forced to give up anatomy, and became the court physician in Madrid. According to some reports, Vesalius was then banished by the Spanish Inquisition, after he was found dissecting the body of a Spanish nobleman, who then woke up during the procedure. As a punishment for this outrage, he was sent on a pilgrimage to Jerusalem, but his ship wrecked, and he died of hunger on the Greek island of Zante.

In truth, Vesalius does not seem to have always been particularly rigorous in making sure his subjects were actually dead. On one occasion, he described removing the still-beating heart of someone who had died in an accident. We might wonder quite what Vesalius's definition of 'dead' was, if a beating heart didn't constitute life.

Why was Edward Jenner rejected by the College of Physicians?

Smallpox is one of the biggest killers in history. No cure has ever been found, and for most of human history the disease consistently killed around 30–50% of all those who become infected. Smallpox is highly contagious, and can be easily transmitted by face-to-face contact. During the 18th century, it was response for one in every ten deaths in Europe, killing around 60 million people. Smallpox is a deeply unpleasant disease. Within days of becoming infected, victims develop hundreds of disfiguring blisters, particularly on their face, neck, and hands. Fever and blindness often follow, with death usually occurring within 10–16 days.

For centuries, people had noticed that anyone who survived one bout of smallpox would then become immune. During the early 18th century, inoculation was introduced to Europe from Turkey, where it had been widely practised for more than a century. Inoculation involves taking pus from a person infected with smallpox, and scratching it into the skin of the person being inoculated. Generally, the recipient will contract a much weakened form of the disease, with no facial scarring, and after that they become immune. However, inoculation was not without risks, as around 2–3% of those inoculated would develop the full form of the disease and die, and some of them would become the source of fresh epidemics. As a result, there was considerable public hostility to inoculation.

Towards the end of the 18th century, a country doctor named Edward Jenner became aware of the local folk wisdom which held that milkmaids in his native Gloucestershire, who were renowned for their beauty and clear complexions, could not catch smallpox. The reason, locals explained, was that most of the girls would generally catch cowpox, from the cows they worked with every day, and they were therefore somehow consequently protected from smallpox. In 1796, Jenner decided to conduct an experiment. He took an eight year boy named James Phipps, and infected him with the pus from a cowpox pustule. The boy fell mildly ill, and then recovered, as was expected. After the boy's recovery, Jenner tried to inoculate him with small-pox, but the infection wouldn't take. Just as Jenner had hoped, Phipps had been rendered immune to smallpox. Jenner named this procedure 'vaccination', after the Latin word 'vacca', which means 'cow'.

After Jenner's discovery was made public, there was a great deal of opposition to vaccination, as it was seen by many as a disgusting and unnatural practice to infect people with a disease from a cow. Jenner's first paper to the Royal Society was rejected, and he had to perform a great number of further experiments before his argument was taken seriously. However, the clear evidence for the effectiveness of vaccination won out, and the practice became widely used. In 1840, the British government banned smallpox inoculation, in a clear statement that vaccination had won the day.

In fact, Jenner did not invent vaccination, as it had been performed before by other country doctors (although it's not clear whether or not Jenner was aware of this). Nonetheless, his success in popularising vaccination, and proving the case for it, mean that Jenner is widely held to have saved the lives of more people than any other man in history. In recognition of his achievement, he was appointed Physician Extraordinary to King George IV – a considerable honour – and feted and celebrated all over the world. Back in the 1680s, somewhere between 7–14% of deaths in London had been attributable to smallpox; by 1850, this had fallen to just 1%.

However, the world of medicine was harder to impress. After his vaccination triumph, Jenner applied to join the College of Physicians in London, but he was denied entry, and told that he would first have to pass a test on the theories of Hippocrates and Galen. Jenner refused to take the test, as he felt that his achievement in defeating smallpox was enough to make him worthy of election. However, this argument was clearly not acceptable to the hidebound dinosaurs of the college, and so Edward Jenner, who may well have saved more lives than any man in history, was never accepted into the College of Physicians.

Why was Hungarian doctor Ignaz Semmelweis hounded out of medicine?

Ignaz Semmelweis worked in the maternity wards of the Vienna General Hospital in the 1840s, and was very troubled by the wards' terrible fatality rates from puerperal fever, also known as 'childbed fever'. The hospital had two obstetrical clinics: one was run by male medical students, while the other employed midwives. It was well known that the death rate in the students' clinic was much higher than that in the midwives' clinic, but no one knew why. Pregnant women would literally beg to be admitted to the midwives' clinic, fearing for their lives. If they were denied, many of them chose to simply give birth in the street, rather than take the risk of being treated in the student clinic, where mortality rates were as high as 15%.

Semmelweis was determined to work out why the mortality rates in the two clinics were so divergent, but he could find no significant difference between the two clinics, apart from the staff they employed. In 1847, Semmelweis's friend Jakob Kolletschka died from a condition similar to puerperal fever, after being accidentally cut by a scalpel while performing an autopsy. This sad event proved to be the inspiration for Semmelweis's great discovery. He realised that the medical students were somehow becoming contaminated by the corpses, while performing autopsies, and were then transferring this contamination to the expectant mothers in the obstretrical clinic, when they went straight from the morgue to the delivery room. Midwives never performed autopsies, which is why the midwives' clinic was not contaminated in this way.

Semmelweis instituted a radical new rule: all medical staff were now required to wash their hands and instruments in chlorinated lime solution in between performing autopsies and dealing with live patients. The results were immediate: the mortality rate in the students' clinic quickly dropped to the same level as that in the midwives' clinic. It was a remarkable success.

However, Semmelweis's discovery proved extremely controversial, for a number of reasons. Firstly, the mainstream medical view at the time was that illnesses were caused by an imbalance in the four humours, which was unique to the patient. This theory held that all medical conditions were individual and specific to the patient, whereas Semmelweis was arguing that one overall, invisible cause was responsible for all the deaths. The second objection was that doctors at the time believed that diseases were spread by 'bad air' or miasmas, which meant there was no possible mechanism for them to be transmitted in tiny, invisible quantities on a person's hands. Perhaps most importantly, Semmelweis's theory was controversial because it carried the implicit suggestion that doctors were responsible for spreading disease, rather than curing it, which was a heretical and dangerous notion. As a result, Semmelweis was derided and criticised by the medical establishment, and hounded out of Vienna.

He moved to Pest (now Budapest), and in 1851 took an unpaid role at the obstetric ward at the small St Rochus hospital. When he arrived, childbed fever was out of control at the hospital, but Semmelweis quickly wiped it out – between 1851 and 1855, less than 1% of the hospital's patients died of the disease. Nonetheless, Semmelweis's consistent record of success was still overlooked by the medical establishment, and his methods were ignored.

He became angry and bitter, writing frequent polemics condemning his critics. As his health deteriorated, and his behaviour became more erratic, he was committed to a lunatic asylum. He died two weeks later, after being severely beaten by the guards and kept in a straitjacket in a darkened cell. He died of blood poisoning, the same condition which in a maternity ward would be called childbed fever. The rules of the Hungarian Association of Physicians required that every member who died should be honoured with a commemorative address, but Semmelweis's death was ignored, and no colleagues attended his funeral.

After his death, Semmelweis's important contribution to medicine became widely recognised, and today he is regarded as a pioneer of antiseptic medicine, and many medical institutions in Austria and Hungary have been named in his honour.

Did an African-American pioneer of blood transfusions die after being refused a blood transfusion, on account of the colour of his skin?

Charles Drew was an American surgeon, born in Washington, DC in 1904. He was a leading researcher in the field of blood transfusions, and in the course of his career he developed new techniques for storing blood, which were key to the development of the large scale blood banks that saved many thousands of lives during World War Two. Drew also took an active role in this project: as the director of the 'Blood For Britain' programme, he supervised the collection of 14,500 pints of blood plasma for the British. Drew also campaigned against the contemporary practice of storing the blood of African-American donors separate from that of white donors; a practice which was based purely on racial prejudice, and had no scientific merit. In 1943, Drew became the first African-American to be elected to serve as an examiner on the American Board of Surgery.

At around 8am on April 1, 1950, Drew was driving back from Tuskegee, Alabama, having spent the night working at the free clinic. Exhausted after his night's work, Drew crashed the car, injuring himself along with the three colleagues who were passengers. The three other doctors suffered minor injuries, while Drew was trapped in the wreckage, with his foot wedged under the brake pedal. When the emergency services arrived, Drew was in shock, with severe leg injuries. He was taken to Alamance General Hospital in Burlington, North Carolina, where he was pronounced dead.

There is a commonly told version of this story, which states that Drew was denied care because of the colour of his skin. Specifically, this version of the story holds that this pioneer of blood transfusions was himself, in a bitterly ironic twist, denied a blood transfusion, and that he consequently bled to death. It is a powerful story, which highlights the shameful practice of seg-regated hospitals that did go on in the South at that time.

However, despite its dramatic power, the story appears to be untrue. According to one of Drew's passengers that night, Dr John Ford, 'We all received the very best of care that night. The doctors started treating us immediately.' In fact, accord-ing to Ford, Drew's injuries were such that a blood transfusion would have been actively harmful. 'He had a superior vena caval syndrome: blood was blocked getting back to his heart from his brain and upper extremities,' Ford said. 'To give him a transfu-sion would have killed him sooner. Even the most heroic efforts couldn't have saved him. I can truthfully say that no efforts were spared in the treatment of Dr. Drew, and, contrary to popular myth, the fact that he was a Negro did not in any way limit the care that was given to him.'

This story is similar to another urban myth, concerning the legendary blues singer Bessie Smith. The story goes that Smith died after a car accident, when a 'Whites Only' hospital refused to treat her. In fact, it seems that the scene of Bessie Smith's accident was served by two ambulances, one from the black hos-pital, and the other from a white hospital. Smith was taken to Clarksdale's Afro-American Hospital, where she died. Although it seems entirely plausible that she might have been refused treat-ment if she had been taken to the white hospital, this is not what happened. Bessie Smith did receive treatment, at the Clarksdale Hospital, including having her right arm amputated, but her injuries were severe, and as a result she died in the hospital.

Who was the real Sherlock Holmes?

Before becoming a writer, Sir Arthur Conan Doyle, the creator of Sherlock Holmes, studied medicine at Edinburgh University. One of his lecturers there was the brilliant Dr Joseph Bell, who would prove to be a pivotal influence on the young Conan Doyle. In his second year, Conan Doyle was chosen to serve as Bell's outpatient clerk at the Edinburgh Royal Infirmary, and as a result was able to observe Bell's methods at close hand. Bell was tall and lean, with piercing green eyes and a large, hooked nose.

Bell was a brilliant diagnostician, who would make amazing deductions about people just by observing their accent, clothing, gait, complexion, and other small, telling details. In this way, he impressed upon his students the importance of close observation in making a diagnosis. On one occasion, a patient entered the room, and Bell immediately stated that this man was a recently discharged non-commissioned officer, who had served in a Highland regiment stationed in Barbados. He explained his deductions, 'You see, gentlemen, the man was a respectful man but did not remove his hat. They do not in the army, but he would have learned civilian ways had he been long discharged. He has an air of authority and is obviously Scottish. As to Barbados, his complaint is elephantiasis, which is West Indian and not British.'

Similar moments of deductive brilliance are of course a recurring feature of Conan Doyle's eternally popular Sherlock Holmes stories. In 'A Study In Scarlet', we see the first meeting between Holmes and his assistant Watson, which has obvious echoes of the scene just described. Upon meeting Watson, Holmes observes, 'You have been in Afghanistan I perceive.'

Watson assumes someone has told Holmes about him, but the detective explains:

> *'Nothing of the sort. I knew you came from Afghanistan. From long habit the train of thoughts ran so swiftly through my mind that I arrived at the conclusion without being conscious of intermediate steps. There were such steps, however. The train of reasoning ran, "Here is a gentleman of a medical type, but with the air of a military man. Clearly an army doctor, then. He has just come from the tropics, for his face is dark, and that is not the natural tint of his skin, for his wrists are fair. He has undergone hardship and sickness, as his haggard face says clearly. His left arm has been injured. He holds it in a stiff and unnatural manner. Where in the tropics could an English army doctor have seen much hardship and got his arm wounded? Clearly in Afghanistan." The whole train of thought did not occupy a second. I then remarked that you came from Afghanistan, and you were astonished.'*

The dull-witted Watson replies, 'It is simple enough as you explain it.'

Correspondence from Conan Doyle confirms that Bell was the inspiration for Sherlock Holmes. After his creation had brought him fame and success, Conan Doyle wrote to his former teacher, 'It is most certainly to you that I owe Sherlock Holmes, and though in the stories I have the advantage of being able to place him in all sorts of dramatic positions, I do not think that his analytical work is in the least an exaggeration of some effects which I have seen you produce in the out-patient ward.'

Bell generously replied, 'You are yourself Sherlock Holmes and well you know it.'

Who discovered antisepsis?

Joseph Lister (1827–1912) was an English surgeon who pioneered the concept of sterile surgery. While working at the Glasgow Royal Infirmary, Lister instigated a policy of antiseptic surgery which involved spraying the operating theatre with carbolic acid, and cleaning the instruments, dressings, and even the wounds themselves with the same solution. Thanks to Lister's sanitary innovations, the rates of post-operative infection and mortality were dramatically reduced, and his ideas were widely adopted.

Although Lister pioneered the idea of antiseptic surgery, various forms of antisepsis had been used since ancient times, before germ theory was ever discovered. Ancient societies would hang meat to dry in the sunlight, or over a slow smoky fire, having learned that this would make the meat resist decay much longer. Ancient Egyptians preserved their mummies using smoke, dry tombs, and embalming chemicals to ward off germs. In a number of early cultures, a common medical treatment was the application of naturally-occurring antiseptic tars and petroleums onto wounds and sores. Around 800 BC, the Hindu surgeon Susrata recommended fumigating operating theatres before and after surgery. He also recommended that water could be kept clean by being kept in copper vessels, filtered through charcoal, and exposed to heat. Nonetheless, although certain antiseptic techniques may have been used before Lister, his reputation as the father of antisepsis is well deserved, as it was he who first applied Pasteur's discoveries of germ theory to the practical improvement of surgery.

Although Lister's approach was revolutionary at the time, if we saw his operating theatre today, it would not strike us as being particularly clean at all. Although he endeavoured to keep the surgical area itself clean, Lister's operating theatres as a whole were not much cleaner than the rest of the hospital. Lister himself performed surgery using the same apron every time, which was so caked in blood that it shone. Today, surgery follows the principle of 'asepsis' rather than 'antisepsis', meaning that rather than simply killing germs during the surgical procedure (antisepsis), the aim is to make the operating theatre and instruments completely free of germs in the first place (asepsis).

What was the 'Guinea Pig Club'?

The Guinea Pig Club was the name given to a group of burns victims who underwent treatment together at the Queen Victoria Hospital in East Grinstead, England during World War Two. Most of the patients were injured fighter pilots and bombers, who had all suffered severe burns, often to their face and hands. The programme was run by Archibald McIndoe, who was later knighted for his efforts. The men became known as the Guinea Pig Club because many of the reconstructive plastic surgery techniques that were used on them were new and experimental. Before the war, any patient suffering from severe burns had only a limited chance of survival, and there were no precedents for many of the procedures that McIndoe carried out.

The Guinea Pig Club group was officially formed in 1941, and consisted of 39 patients. To join the club, you had to have gone through at least ten surgical procedures. Many of the victims were badly disfigured. Air Gunner Les Wilkins had lost his face and hands, and McIndoe had to make incisions between his knuckles to recreate fingers.

McIndoe is today regarded as a pioneer in the field of plastic surgery, as his innovations and inventions marked a huge advance in the subject. However, the Guinea Pig Club was also notable for McIndoe's unusual approach to patient care and rehabilitation. Knowing that many of his patients would be at the hospital for years, he tried to make the conditions as relaxed and communal as possible. Patients were not required to wear the standard uniform of 'convalescent blues', and they could leave the hospital at any time. Barrels of beer were kept in the wards, and McIndoe encouraged local families to take in patients as guests. Burns victims are often badly disfigured, and dealing with the reactions of strangers is one of the biggest challenges, but East Grinstead became known as 'the town that didn't stare'.

The sense of humour and solidarity that was fostered is evident in lots of ways. The club was called The Guinea Pig Club, in cheerful recognition of the fact that the patients were guinea pigs for new and experimental techniques. When appointing board members, the club chose a secretary who had suffered badly burnt fingers, so that the minutes of meetings would not be too verbose. The club's first treasurer was an airman who had severe burns to his legs, which meant he would not be able to run away with the club's funds. There is even a light-hearted club song, with the following first verse:

We are McIndoe's army,
We are his Guinea Pigs.
With dermatomes and pedicles,
Glass eyes, false teeth and wigs.
And when we get our discharge
We'll shout with all our might:
'Per ardua ad astra'
We'd rather drink than fight!

After the war ended, the group continued to meet socially, and by this time it comprised over 600 members. Up until 2007, there were regular get-togethers once a year, but because of the age and frailty of the surviving members, it seems unlikely that any more meetings will take place.

How did a single Soho water pump change the course of modern medicine?

Cholera was the scourge of the nineteenth century. It was initially endemic to India, where it was thought to be spread by the contaminated Ganges river, but from 1816 the disease began to expand its range, travelling along trade routes across Asia. This first cholera pandemic lasted a decade, until 1826, and it almost reached Europe, but receded just in time. The second pandemic started in 1829 and spread across Asia, before reaching Egypt, North Africa, Russia, and then finally reaching Western Europe. Further pandemics continued throughout the 19th century and beyond, causing an estimated 38 million deaths in India alone.

Cholera was a horrific way to die. Sufferers would develop acute nausea, which would progress to violent vomiting and diarrhoea. Their stools would become a grey liquid, described as 'rice water', which soon consisted of nothing more than liquid and fragments of gut. This would lead to extreme, painful cramps, and an insatiable thirst. As the patient neared death, he would develop the classic sign of cholera: pinched blue lips in a shrivelled, sunken face.

No one knew what caused the disease, or how it spread. In London, there had been a number of epidemics since 1832, resulting in tens of thousands of deaths. In 1854, there was a severe outbreak of cholera in a small area of Soho, around Broad Street (which is now called Broadwick Street). John Snow was a doctor based in London, who was sceptical of the prevailing 'miasma' theory, which held that diseases were spread by 'bad air'. Snow studied the Soho outbreak, and noticed that the vast majority of cases had occurred in homes situated near a single water pump on Broad Street. There were only ten deaths in homes that were closer to a different water pump, but Snow learned that most of the victims in these homes actually took their water from the Broad Street pump, either because they preferred the taste, or because they were children who went to a school near the Broad Street pump.

Snow examined the pump and its water closely, but couldn't find any obvious source of the disease. However, although he couldn't physically identify the specific cause of the outbreak, his statistical and geographical analysis was enough to convince the local council to disable the pump, which they did simply by removing the handle, and this measure alone succeeded in ending the cholera outbreak. It was later discovered that the well supplying the Broad Street pump had been dug just three feet from a leaking cesspit, which had become infected with cholera.

Snow's intervention had powerfully demonstrated that diseases could be transmitted through faecally contaminated water, rather than through miasmas of 'bad air'. Although his findings were not immediately accepted by the scientific community, Snow's analysis of the Broad Street outbreak is now regarded as the defining event in the science of epidemiology, which is the study of causes, distribution, and control of diseases in a population. Today, there is still a replica water pump on Broadwick Street to mark Snow's achievement, just yards from the pub named in his honour.

What was unusual about Dr James Barry?

James Barry was a leading surgeon in the British Army in the early 19th century. He graduated from Edinburgh University, and served in Canada, India, and South Africa, reaching the rank of Inspector General. He was known as a passionate reformer, who helped to improve the lives of soldiers and civilians alike. He campaigned for better food and care for lepers and prisoners, and in 1826 became the first British surgeon to perform a successful Caesarean section in which both mother and child survived. Barry was a fiery character, who fought a number of duels, and was arrested and disciplined on a number of occasions. Florence Nightingale described Barry as 'the most hardened creature I ever met throughout the army,' after they clashed in Corfu during the Crimean War.

In 1865, Barry fell victim to an epidemic of dysentery, and died in London. While preparing the body for burial, house-maid Sophia Bishop made an astonishing discovery. According to Bishop, Barry was actually 'a perfect woman!' Not only that, but a woman who had given birth, judging by the stretch marks Bishop observed on the body. This meant that Barry had in fact been the first woman to qualify as a doctor in Britain, and that she had successfully hidden her sex from her staff and colleagues for almost 50 years.

However, it seems that some of them may have had suspicions. When posted to South Africa in 1824, Barry was seen to have a close relationship with the colony's governor, Lord Somerset, which even led to the accusation of a homosexual relationship. Barry's colleagues had also noticed her smooth chin, small frame, and high-pitched voice. She wore padded shoulders, and shoes with lifts. Lord Albemarle remarked that Barry 'showed a certain effeminacy in his manner'. After Barry's death, when the truth became public, a number of Barry's colleagues claimed to have known all along.

So how did she manage to disguise herself for so long? The answer was recently discovered in a collection of family correspondence, which confirmed that Barry's family had helped her to pull off an elaborate conspiracy, which began before her acceptance to Edinburgh University. Barry was born in Cork some time between 1792 and 1795 as Margaret Bulkley. When her father died, the family were left destitute. Margaret and her mother moved to London, and came up with a plan that would allow Margaret to fulfil her obvious potential, while financially supporting the pair of them. At this time, of course, only men were allowed to study medicine and become doctors. However, thanks to a number of influential friends, Margaret began to study medicine privately. In 1809, Margaret and her mother set off on a ship bound for Scotland. Before boarding, Margaret changed her identity, disguising herself as a man. She then presented herself as James Barry throughout the voyage, and thus avoided any risk of discovery on her arrival. From that point on, she seems to have lived her life as a man, and escaped discovery until her death.

Was a derided female scientist the true discoverer of the structure of DNA?

The scientist in question was a lady called Rosalind Franklin, and although there is still considerable debate about precisely how much of the credit she deserves for her part in this momentous breakthrough, few would dispute that she played a crucial and undervalued role.

Franklin was a brilliant biologist and chemist, who had learned X-Ray diffraction techniques at the Laboratoire Central Des Services Chimiques in Paris. In 1951, she joined the department at King's College, London, which at that time had a patronising and sexist culture towards women. Only men were allowed in the university dining rooms, and after work Franklin's colleagues would go to relax in men-only pubs. At King's, Franklin was assigned to lead a research project into the structure of DNA. Her colleague Maurice Wilkins was working on a similar, but separate, project. The two of them were peers, but Wilkins automatically assumed that Franklin was his technical assistant, which made their relationship frosty from the start.

By this time, much was already known about DNA and its role as a store of genetic material. What was not known was what the molecules looked like, or how they functioned. Franklin used pioneering crystallographic X-Ray photography to take incredible pictures of the intricate, double-helix of DNA, pictures which have been described as, 'the most beautiful X-ray photographs of any substance ever taken.'

Without Franklin's knowledge, Wilkins showed Franklin's work to James Watson, who was pursuing similar research into

DNA at Cambridge University. Franklin's research, and in particular her photographs, providing Watson with the crucial data he needed to complete his own model of DNA. Watson quickly published his model in the scientific journal Nature, naming himself and his colleague Francis Crick as authors, with only a vague footnote mentioning Franklin and Wilkins' work. The research carried out by Wilkins and Franklin was also included separately, but presented in a way that added to the impression that Franklin's research had been of only minor significance, as Crick and Watson's article came first, followed by Wilkins', with Franklin's paper third in line.

Franklin left Kings, and continued to carry out groundbreaking research, into the tobacco mosaic virus, and then polio. In 1956, she became ill with cancer, and died. It is possible that her cancer may have been caused by exposure to radiation while working on the X-ray techniques that revealed the structure of DNA. After her death, James Watson released a book called 'The Double Helix', in which Franklin was mentioned only fleetingly, and in disparaging terms. In the book, Watson describes Franklin as being Wilkins' assistant, and suggests that she was incapable of interpreting her own data. In 1962, when Watson, Crick and Wilkins were awarded the Nobel Prize, Franklin wasn't even mentioned.

Today, Rosalind Franklin is increasingly recognised for her contribution. Her portrait at the National Portrait Gallery hangs alongside those of Watson, Crick, and Wilkins, and in 1992, English Heritage erected a blue plaque at her childhood home. In 2003, the Royal Society established the Rosalind Franklin Award, to reward outstanding contributions to any area of natural science, engineering, or technology. In 2009, the readers of New Scientist magazine voted Franklin second only to Marie Curie in a poll to find the most inspirational female scientist.

Why did Baron Dupuytren remove fat from corpses?

Baron Guillaume Dupuytren (1777–1835) was a leading French surgeon, who was celebrated for his skill and dexterity. He treated Napoleon Bonaparte's haemorrhoids, and he was one of the first surgeons to successfully drain a brain abscess using trepanation, a process by which a hole is made in the skull. He was a prolific writer, and over the course of his life he amassed a great fortune.

He is best remembered today in connection with a condition called Dupuytren's Contracture, in which the fingers bend inwards towards the palm and cannot be straightened. Dupuytren was the first person to devise and perform an operation to correct this condition, which is ironic given that he himself was so tight-fisted. According to legend, when he was a medical student, Dupuytren would take fat from the corpses in the dissection room to burn in his reading lamp at home. Dupuytren was widely admired, but equally disliked, as he was seen to be haughty, unfriendly, and obsessive.

On one occasion, he had saved the life of a duchess. The grateful lady presented him with a beautiful, hand-embroidered purse as a token of her gratitude, but Dupuytren snapped back that his fee was actually 5,000 francs. The duchess took back the purse, opened it, and removed from it five 1,000 franc notes. She then handed the purse back to him, telling him that it now contained exactly the amount for which he'd asked. Smiling, she complimented him on his modest fees.

CHAPTER SIX
BAD MEDICINE

'Never go to a doctor whose office plants have died.'

Erma Bombeck

Was Jack The Ripper a surgeon?

Jack The Ripper was one of the world's most infamous serial killers, whose unsolved series of murders make up one of the most mysterious and fascinating cases in the annals of crime. He – or according to some, she – is generally believed to have committed somewhere between five and eleven murders between 1888 and 1891, in the impoverished slums of Whitechapel in the East End. However, there is considerable debate about almost every detail of these crimes.

If there can be said to be a consensus, it is that Jack The Ripper committed five murders, known as the 'canonical five', between August and November 1888. The victims were Mary Ann Nichols (who was killed on the 31st of August), Annie Chapman (8th September), Elizabeth Stride and Catherine Eddoweson (both 30th September), and Mary Jane Kelly (9th November). All five women worked as prostitutes, and they were all killed by having their throat cut, before being horribly mutilated (with one exception, Elizabeth Stride, possibly because Jack was interrupted during the crime and fled).

The murders remain unsolved, despite one of the largest police investigations in history, as well as the ongoing efforts of an army of journalists, academics, and amateur investigators, who have turned 'Ripperology' into a field of study all of its own. So far, more than a hundred people have been identified by Ripperologists as potential suspects, even though most of them were never considered credible by the police at the time, and there are a whole range of fascinating conspiracy theories.

However, for the purposes of this question, the interesting detail concerns the Ripper's medical knowledge. At the time many experts, including the police, were convinced that the Ripper must have been a butcher, an abattoir worker, or a doctor of some kind, because of the speed and skill with which he attacked and dismembered his victims. Four of the five victims had their abdomen cut open. The killer removed the uterus of Annie Chapman, the heart of Mary Jane Kelly, and the uterus and left kidney of Catherine Eddoweson. Half of this kidney was later posted to the police, inside one of a series of taunting letters, with the sender claiming to have eaten the other half. Witnesses stated that they saw a 'shabby, genteel man' carrying a shiny black bag at the scene of one of the murders, which added to the suspicion that the killer was a medical man. The police interviewed all the local butchers and abattoir workers, but found they all had alibis, eliminating them from the investigation. As a result, a number of doctors have long been considered key suspects. They include the following:

Dr Francis J. Tumblety was a 56 year old American quack, who had toured the US and Canada selling 'Indian' herbs, and was arrested for his involvement in the assassination of Abraham Lincoln, but never charged. Tumblety was in England in 1888, and he became one of the Metropolitan Police's four main suspects for the Ripper murders. He was arrested in November 1888 for 'offences of gross indecency', which in this case seems to have meant homosexuality. He fled the country, and his flight coincided with the end of the killings, which adds to the case against him.

Michael Ostrog was another of the police's four central suspects, Ostrog was a Russian-born con artist. He had a record of fraud and petty theft, but nothing particularly serious or violent. However, he did claim to have been a surgeon in the Russian Army.

George Chapman was born in Poland as Seweryn Kłosowski, before changing his name when he moved to England around 1887–8. Chapman certainly was a murderer, as he poisoned three of his wives; crimes for which he was hanged in 1903. He was known to be a violent misogynist, and he was also a trained surgeon. However, it's very unusual for a single serial killer to kill using two such distinct methods, and there's little meaningful evidence linking Chapman to the specific Ripper murders.

Thomas Neill Cream was a Scottish doctor, who specialised in abortions. Cream is known to have committed at least five murders, and most of his victims were prostitutes, whom he murdered by poisoning. He was thought to have been in prison at the time of the Ripper murders, but some believe he may have bribed his way out, or even arranged for a lookalike to take his place. When he was hanged in 1892, his last words on the gallows are said to have been, 'I am Jack The ...'

As the years have passed by, and the facts have become ever more distant, increasingly outrageous suspects have been suggested. They include Lewis Carroll, the writer of 'Alice's Adventures in Wonderland'; Sir William Gull, physician to Queen Victoria; Prince Albert Victor, the Queen's grandson; Sir John Williams, obstetrician to Queen's daughter Princess Beatrice; and the noted artist Walter Sickert. Given the amount of time that has passed, and the efforts that have already been made, this mystery will surely never be solved, but it still makes for a fascinating puzzle.

Which renowned surgeon accidentally cut off his patient's testicle?

Robert Liston was a leading surgeon of the 19th century, who was famous for the extraordinary skill and speed with which he could perform complex surgerical procedures. In an era before anaesthetics, the speed of an operation was arguably the most important factor in terms of preventing pain and increasing the patient's chances of survival. It was said that Liston could amputate a leg, and sew up the end, in just 90 seconds. Liston was tall and macho; he would stride into the operating theatre, instructing those in the spectators' gallery to time him with their pocketwatches, as he set to work. To keep his hands free, he would hold the bloody knife between his teeth. According to his admirers, 'The gleam of his knife was followed so instantaneously by the sounds of sawing as to make the two actions appear almost simultaneous.'

However, this emphasis on speed could sometimes come at the expense of accuracy. On one occasion, Liston amputated the patient's leg in just two and a half minutes, but he also accidentally sliced off the patient's left testicle. The audience at another amputation witnessed Liston cut off two of his assistant's fingers, as well as the coattails of a distinguished spectator. The spectator dropped dead on the spot, from sheer terror. Later, both the patient and Lister's assistant died from gangrene, as a result of their injuries. This extraordinary event has thus been memorably described by the great medical historian Richard Gordon as, 'The only operation in history with a 300 percent mortality.'

What were 'ether frolics'?

The history of anaesthesia is a rather strange story. When they were first discovered in the 18th and 19th centuries, the first effective anaesthetics were used solely for entertainment, and it barely seemed to occur to anyone to use them to alleviate the horrific pain of surgery.

Before this time, numerous methods of pain relief had been tried, with only limited success. In ancient Egypt, patients would be hit on the head with a mallet. Later, Europe's doctors spent centuries experimenting with ice, snow, hypnotism, suffocation, sleeping draughts, opium, wine, and cannabis. One doctor whom we have already encountered, the famously tight-fisted Baron Dupuytren, developed the innovation of shocking his female patients with crude insults, in the hope of making them faint. However, none of these solutions was satisfactory. The only method that actually worked was drugs, but the problem with drugs, as the great Italian physician Fallopius complained, was that, 'When soporifics are weak, they are useless, and when strong, they kill.' Because of this inherent danger, a 17th century French law had banned the use of any painkillers during surgery.

Nitrous oxide, which is also known as laughing gas, was discovered by Joseph Priestly in 1772, but it was used primarily to create entertaining stage shows, in which members of the audience would come on stage, inhale a dose of nitrous oxide from an Indian-rubber bag, and then begin laughing, dancing, tumbling, and generally going berserk for the amusement of the crowd. One man, a dentist named Horace Wells, did notice the drug's painkilling potential, but after one unsuccessful public demonstration he gave up and lost interest.

Ether was first synthesised in 1540, and its stupefying powers were well known. By the early 19th century, one of the more popular pastimes in fashionable society was to hold 'ether frolics', which were parties fuelled not by champagne and cocktails, but by ether. The parties were a hit, but it didn't seem to occur to anyone to use ether for pain relief until the 1840s, when an

American doctor called Crawford Long used it to painlessly remove a cyst from a friend's neck.

Even after anaesthetics had been convincingly proven to work, there was still considerable hostility to their use. Ether was referred to by surgeons as the 'Yankee Dodge', as doctors seemed to feel that avoiding pain was somehow unsporting, like a form of cheating. Perhaps their professional pride was stung. In one stroke, anaesthesia had made much of the surgeon's craft useless, as their speed, skill, and indifference to pain were no longer of any value. However, although the doctors were unconvinced about the merits of anaesthesia, the public were of one mind. The People's London Journal ran the headline, 'HAIL HAPPY HOUR! WE HAVE CONQUERED PAIN!', and for a brief time 'Anaesthesia' became a fashionable name for newborn girls.

What did Cezanne, Monet, and Van Gogh have in common?

Obviously, one answer is that they were three of the most important painters of the Impressionist/post-Impressionist period, but fine art falls outside the scope of this book. For the purposes of this book, the more relevant connection is that they may all have suffered from the same ailment.

The likely cause was a type of paint, specifically a pigment called Emerald Green, which was very popular among artists of the time. Emerald Green is made from copper acetate and arsenic trioxide, and it is the arsenic in Emerald Green which is the issue, as arsenic is of course extremely toxic. Emerald Green is no longer used by painters today, but it's said that none of our modern equivalents is quite as vivid and brilliant as Emerald Green.

Emerald Green is said to have been Cezanne's favourite pigment, and his paintings are full of it. Van Gogh and Monet also used it heavily, and all three men suffered from symptoms of chronic arsenic poisoning: Cezanne developed severe diabetes,

Van Gogh suffered from neurological disorders, and Monet went blind. All three may well have been caused by the artists' use of Emerald Green.

On the other hand, a number of other toxic pigments might also have been the cause. Another popular pigment was a purple called Cobalt Violet, which also contained arsenic. Other commonly used pigments contained lead, mercury, and turpentine, which are all poisonous. One way or another, it does seem likely that all three men really did suffer for their art.

Was King George V killed by his doctor?

The final years of King George V's life were marred by illness, thanks in part to his heavy smoking. He suffered from emphysema, pleurisy, and bronchitis, and on January 15th, 1936 he took to his bed at Sandringham House, Norfolk, complaining of a cold. His condition swiftly deteriorated, and by January 20th he was close to death. His doctor Lord Dawson issued a brief statement, stating simply, 'The King's life is moving peacefully towards its close.' At 11.55pm that night, King George passed away.

However, George did not die of what are euphemistically known as 'natural causes'. The cause of death was a deliberate, fatal injection of drugs: three-quarters of a gram of morphine, and one gram of cocaine, administered by Lord Dawson himself. Incidentally, this combination of drugs is what's known in certain circles as a speedball, and it's what killed River Phoenix and John Belushi, albeit accidentally in their cases.

So why did Dawson kill the King? There seem to have been two reasons. Firstly, the King was clearly dying, and seeing him struggle through his final hours was felt to be beneath his dignity, and distressing for his family. Years later, Lord Dawson explained:

At about 11 o'clock it was evident that the last stage might endure for many hours, unknown to the patient but little comporting with the dignity and serenity which he so richly merited and which demanded a brief final scene. Hours of waiting just for the mechanical end when all that is really life has departed only exhausts the onlookers and keeps them so strained that they cannot avail themselves of the solace of thought, communion or prayer.

The other, astonishing reason was that the Royal Family were anxious that the news of the King's death should appear first in the morning edition of The Times newspaper, rather than the 'less appropriate' evening papers, or even the upstart BBC wireless. Therefore, it was decided that the King should die in time to meet The Times' print deadline. Dawson had even prewarned the editor to hold the front page.

Of course, none of this became public knowledge at the time. In fact, it only came out fifty years later, in 1986, when Lord Dawson's notes were released to the public. There was surprisingly little public outcry, no doubt largely because of the length of time that had passed. A spokesman for Buckingham Palace would only say, 'It happened a long time ago, and all those concerned are now dead.' However, there were a few outraged voices. George's biographer Kenneth Rose was appalled, stating that, 'In my opinion the King was murdered by Dawson.'

Presumably, doctors never harmed any other kings though, did they?

I'm afraid they did, although in most of the cases that follow, the damage tended to result from well-meaning incompetence, rather than deliberate euthanasia. King Charles II of England, for example, was effectively tortured to death by his doctors, who spent the last four days of his life trying every kind of barbaric treatment without success. He was repeatedly bled, purged, clystered, cupped, blistered, and sweated by his baffled physicians. In the end he died of apoplexy, an outdated term which essentially means blood loss and loss of consciousness, the very things his doctors were causing, so it was almost certainly his treatment that killed him, rather than any underlying illness.

King George III was also poorly treated by his doctors. George suffered in the last years of his life from regular bouts of mental illness (inspiring the film 'The Madness of King George') which it is now thought may have been caused by a blood disease called porphyria. By 1810, George was dangerously ill, practically blind, and in considerable pain. The standard treatment at the time for someone with George's symptoms of mental illness was to force the patient into a straitjacket and so, very respectfully, this is what the king's doctors did.

In June 323 BC, Alexander The Great developed a fever, after going on a two-day drinking bender. Within twelve days he was dead, having gradually lost the power of speech. There are various theories about his death, including that he was assassinated by deliberate poisoning, but the most likely explanation is that he died of an inadvertent overdose of hellebore, a herbal medicine prescribed by his doctors. His last words were said to have been, 'I die by the help of too many physicians.'

George Washington was also almost certainly killed by his doctors. He fell ill on Friday 13th December, after having ridden around his farm in the rain the night before, and then eaten dinner while still in his wet clothes. The next day, he felt even worse, so he asked his plantation manager to take some of his blood; the man took about a third of a litre. During the next few hours, two doctors arrived, and each of them bled Washington – in total, he probably lost more than half his blood in a single day. He was also purged, vomited, and blistered, all of which would have dehydrated and weakened him further. By the end of the day, instead of flowing normally as it should have, Washington's remaining blood was thick and sticky, and oozed slowly out of his veins. He died that evening, having pleaded with his doctors, 'Pray take no more trouble about me. Let me go quietly.'

President James A. Garfield was also probably killed by his doctors. He was shot by assassin Charles J. Guiteau on July 2nd 1881, while waiting for a train. In those days, it was not routine for presidents to have bodyguards. During an eleven-week vigil, his condition fluctuated, until his eventual death on September 19th. Although Garfield was shot more than once, many believe that he would have survived had it not been for his incompetent doctors, who repeatedly probed his wounds with unsterilized fingers, trying to locate the bullet that was lodged inside him. One doctor punctured the President's liver in doing so. As a result, Garfield soon developed blood poisoning and infection, losing more than 60 pounds, before eventually dying of a heart attack.

Do patients ever wake up during surgery?

Horrifyingly, this does happen. 'Anaesthesia awareness' is the proper term, and it is currently one of the major talking points in the world of anaesthesia. When a patient is put under general anaesthetic, they are usually given two separate, different drugs. The first is a general anaesthetic, which makes them unconscious and insensitive to pain. The second is a muscle relaxant, such as curare, which paralyses their muscles, making them physically unable to move or communicate in any way.

However, if the equipment delivering the anaesthetic is faulty, or if the anaesthetist makes a mistake, the patient may regain consciousness and feel pain, but be unable to communicate this to the medical team in any way, because their entire body is paralysed by the muscle relaxant. This can occur for a number of other reasons apart from equipment malfunctions and human error. Some patients become more resistant to anaesthetics, particularly if they are long-term users of tobacco, alcohol, or opiate drugs such as heroin.

The problem is exacerbated by the precise nature of anaesthesia. The risk of side effects increases depending on how much anaesthetic is given, so anaesthetists have to tread a fine line, providing just enough anaesthetic to prevent consciousness, while minimising the risk of complications which comes with providing too much anaesthetic.

According to estimates, anaesthesia awareness occurs in around 0.1–0.2% of operations. Only around 42% of these patients will feel actual pain; the rest will be conscious, but pain-free. Even so, the vast majority of patients who become conscious will panic and become anxious as a result, and many develop lasting psychological effects, including Post-Traumatic Stress Disorder.

In recent years, scientists have developed a number of techniques and technologies to try to address this important problem. These include a number of types of brain wave monitor, which monitor electrical activity in the brain. In theory, the rate of electrical activity in the brain is distinctly reduced when a person is under general anaesthetic, but none of these systems seems to be perfect at present, largely because some anaesthetics do not reduce the electrical activity rate in this way.

Another potential solution is the 'isolated forearm technique', which has been pioneered by Professor Wang and Dr Ian Russell at Hull Royal Infirmary. The technique works as follows. A tourniquet is applied to the forearm, cutting off the blood supply, just before the muscle relaxant is administered. As a result, if at any point the patient regains consciousness, and wants to communicate this to the medical team, he or she will be able to move their hand.

Bizarrely, an alternative solution may simply be to deliberately keep more patients awake. In recent years, doctors have conducted a number of operations under what is known as 'light anaesthesia'. This technique involves an epidural injection into the spine, which blocks pain, but allows the patient to stay awake, and breathe and talk normally. Using light anaesthesia, doctors have been able to carry out major operations, including heart bypass surgery. Light anaesthesia not only averts the risk of anaesthesia awareness (as patients can simply tell the doctor if they find they are in pain), it also avoids some of the risks and side effects associated with general anaesthesia. There are other benefits too. Light anaesthesia generally allows patients to go home earlier, increasing patient satisfaction, and cutting costs.

Finally, did you know that people with red hair need more anaesthesia than blondes and brunettes? Amazingly, there seems to be a link between one hormone which affects skin pigmentation, and another which relieves pain. As a result, natural redheads are more sensitive to certain types of pain, and consequently require around 20% more anaesthetic to reach an equivalent level of insensitivity!

Was Abraham Lincoln poisoned?

During the US Civil War, Abraham Lincoln was noted for him calmness and composure under pressure. However, only a few years earlier, he had been regarded as a violent, volatile man, prone to flying into verbal and physical rages. During a debate in 1858, he grabbed a former aide and shook him 'until his teeth chattered'. According to his partner at a law firm, Lincoln could get 'so angry that he looked like Lucifer in an uncontrollable rage.' So what had changed?

The answer may be that Lincoln was a long-term victim of self-inflicted mercury poisoning. During the 1850s, Lincoln suffered from what he described as 'melancholy', which sounds like it might been what we would today call clinical depression; one contemporary described him as living in a 'cave of gloom'. To remedy this, he took a drug called Blue Mass, which was widely used in the 19th century for a whole range of conditions, including depression, toothache, constipation, and the pain of childbirth.

Scientists at the Royal Society of Chemistry have recently tested genuine samples of Blue Mass pills from the period, and found that their main effective ingredient was mercury, which is of course highly toxic. The pills contained more than 120 times what is now considered to be a tolerable daily dose of mercury. The symptoms of mercury poisoning include nausea, diarrhoea, vomiting, and dehydration. It can also cause insomnia, tremors, and outbursts of rage, all of which Lincoln is reported to have suffered from.

Soon after his inauguration in 1861, Lincoln stopped taking the pills. In a letter to a friend, he explained that they 'made him cross'. The symptoms of mercury poisoning are reversible, which may well explain how the violent, temperamental Lincoln became the calm, unruffled statesman of the US Civil War.

Which popular drug may save the rhino?

The answer is Viagra. For thousands of years, powdered rhinoceros horn has been used as an aphrodisiac in traditional Chinese medicine. Of course, powdered rhino horn doesn't actually work as an aphrodisiac, but that hasn't stopped people from killing huge numbers of rhinoceroses to pursue this deluded belief.

Like much of 'traditional' medicine, the logic behind powdered rhino horn as an aphrodisiac is based on superstition rather than science. A rhinoceros's horn is rigid, curved and upright, and thus it vaguely resembles an erect penis. Therefore, through the concept of 'sympathetic magic', a principle similar to ancient Western medicine's Doctrine of Signatures, it was thought that rhinoceros horn must therefore be an effective cure for impotence and erectile dysfunction. By a similar logic, tiger penis is another highly prized aphrodisiac in traditional Eastern medicine, because of the tiger's obvious virility and aggression. As a result, tigers too have been hunted to the brink of extinction.

The irony is that the growth markets for 'traditional' medicine today are not India, China and the East, but rather the pampered West, where we have enjoyed the fruits of advanced, effective medicine for long enough to take them for granted. In China, people are increasingly demanding vaccines, drugs, and real, effective medicines, all the benefits of the advanced scientific system which has led to dramatically increased life expectancies in the West. And so when it comes to improving their sex lives, the Chinese increasingly seek out Viagra, Cialis, and other drugs that actually work, rather than powdered rhinoceros horn. Hopefully, this trend, along with the various laws which have been introduced to protect these animals, will help to stimulate the fecundity of the rhinos themselves, and allow their numbers to increase.

What is a 'paraffinoma'?

We tend to think of cosmetic procedures like botox and collagen injections as being recent innovations, the product of today's shallow, vacuous culture, obsessed with youth and beauty. However, procedures of this type are far from new. As far back as the 19th century, doctors were experimenting with similar injections, for exactly the same purposes. In particular, there was a brief trend for paraffin wax injections, which were designed to smooth out wrinkles, just as botox and collagen do today. Paraffin wax was also injected into breast tissue, in an early form of breast augmentation surgery.

However, there is a very good reason why we no longer hear about paraffin wax injections, as these procedures generally had disastrous results. The injection of paraffin wax causes paraffinomas, which are hard, painful lumps that develop into unsightly scars, and can become infected.

Amazingly, despite this, the practice didn't die out completely, and there have been occasional reports of paraffin wax injections throughout the 20th century. In particular, paraffin wax is sometimes used for penis enlargement. One report from 1956 describes a patient suffering from premature ejaculation, who turned to paraffin wax after other treatments had failed. The procedure was not successful. Six years after the injection, his premature ejaculation had not been cured, and he now suffered considerable pain whenever he got an erection. Eventually the patient had the paraffinoma removed surgically, along with the skin of his penis, which was replaced with skin taken from his scrotum. This surgery was successful, although it left him with a smaller scrotum than before.

Should babies sleep on their front or their back?

One of the most influential books on childcare was 'Baby and Child Care' by the American paediatrician Dr Benjamin Spock. The book was first published in 1946, and went on to become a record-breaking bestseller, selling more than 50 million copies worldwide, and having a profound influence on generations of new parents. The book is still widely praised for its sensible and liberal approach, but one of its recommendations has gone dramatically out of fashion.

In the book, Spock recommends that babies should be put to sleep on their front, which is called the 'prone' position. His rationale is that if a baby vomits in the night, there is more risk of choking on the vomit if the infant is on its back (which is known as the 'supine' position). Thanks to Spock's book, this approach became the orthodoxy, and the vast majority of American babies were put to sleep on their front.

However, it now seems that this advice was flawed. Studies into cot death, which is also known as Sudden Infant Death Syndrome (SIDS), have found that sleeping in the prone position increases the risk of cot death, although it's not clear why. There are various theories. A baby sleeping on its front may be more

more likely to repeatedly breathe in the same, recycled air, which is consequently higher in carbon dioxide. Alternatively, a baby sleeping on its front may have a greater risk of suffocating itself. A third theory is that a baby sleeping on its front may be more at risk of inhaling toxins or mould that could be present in the mattress.

Although the mechanism is unclear, the statistics are compelling. In 1994 a campaign called 'Back To Sleep' was launched to persuade parents to put their babies to sleep on their backs. Since then, the rate of supine baby sleep in the US has gone up from 13% to 76%, while the rate of SIDS has plummeted by more than 50%. While this is good news, the alarming implication is that Dr Spock's well-meaning advice may have led to tens of thousands of unnecessary deaths.

On a lighter note, the Star Trek character Dr Spock was apparently not named after Dr Benjamin Spock; it is just an odd, but perfectly logical, coincidence. According to creator Gene Roddenberry, when he was developing his original proposal for a TV show called Star Trek, he was just looking for an alien-sounding name, and had never even heard of the paediatrician.

Do surgical implements ever get left
inside patients after surgery?

Worryingly, they do, and it's possible that this happens quite frequently. Doctors and nurses are often overworked and sleep-deprived, and they have to perform emergency surgical procedures in a hurry. During a routine operation, on average around 250–300 different tools and implements are used, and this number can be as high as 600 in complicated procedures. As a result, it's not entirely surprising that some things occasionally get lost.

A wide range of tools and instruments have been found inside patients, including needles, forceps, knife blades, tweezers, safety pins, scalpels, clamps, towels, and scissors. One patient had to have a 9.5 inch spatula removed after an operation. Another patient recently had a two-foot guide wire removed, which ran all the way from his groin to his upper chest. Apparently, after the initial operation, doctors inspected the patient's X-rays on six separate occasions without noticing the wire. One surgeon even left a set of his own forceps inside a patient, which were engraved with his initials.

However, the most common objects left inside patients are gauzes, which are also known as sponges. These are small pads of sterile cotton, which one might imagine would be relatively harmless. However, they can in fact cause considerable harm. After surgery, gauzes left in the body can be mistaken for abscesses or tumours, leading to further risky and unnecessary surgery.

One report estimates that there are around 2,700 incidents like this each year in the United States, but this number may well be an underestimate, as many patients are not aware of what has happened, and hospitals are understandably wary of encouraging expensive lawsuits. Patients undergoing emergency surgery face a greater risk of an item being lost, as do obese patients. According to research, a patient with a one-point higher body mass index faces a 10% higher risk of having some implement left inside them.

Hospitals are experimenting with a number of techniques and new technologies to address this problem, but so far none of them seems to be foolproof. It is generally standard procedure to conduct an inventory of every tool and implement a number of times at different stages during an operation, and some hospitals recommend as many as four separate counts. However, there are lots of occasions when there is simply no time for this, and corners have to be cut. It's also not clear how much the counts even help. According to studies, even when all the counts do take place, mistakes still get made. In many cases where an implement has later been found inside the patient, the count that took place after the surgery found nothing missing.

There are also a number of technological solutions being developed. One of them involves marking every gauze with 'radio-opaque' marking, which will then show up on standard radiographic scans. Another potential solution involves marking every single gauze with an individual barcode, which is then scanned into a reader at the end of the surgery.

Which doctor drank a glass of cholera, and survived?

Despite John Snow's triumph with the Broad Street water pump in 1854, there was still considerable debate for decades afterwards about what caused cholera, and how it was transmitted. By 1892, the German doctor Robert Koch had succeeded in isolating the bacterium responsible for cholera, and was advocating a 'germ theory' of disease, which proposed that many diseases were caused by micro-organisms, rather than miasmas, bad air, or imbalances of humours in the body.

However, Koch had not yet done enough to convince the world, and there were many rival theories. Max von Pettenkofer preferred the 'ground water theory', which proposed that rotting organic material in the soil was what released cholera into the air, and once released, the cholera would only infect those who were particularly susceptible, as a result of their poor diet or weak constitution. In other words, the cholera bacterium was not the important factor – the important, decisive factors were the atmosphere, and the intrinsic health of the patient. Therefore, von Pettenkofer believed that cholera could not be transmitted from person to person, but only via the atmosphere, and only to those people who were susceptible.

So confident was von Pettenkofer about his assertions, that in front of a number of witnesses, he lifted a beaker of water which contained a growth of Koch's cholera germs, and drank it down. Knowing what we do today, this seems like an absurd risk to take. However, von Pettenkofer suffered barely any ill effects, apart from a bout of diarrhoea, and consequently he felt that in this one audacious act he had conclusively disproved Koch's germ theory. The rest of the world largely concurred, and so this single moment of foolishness set back the germ theory for decades.

However, we know now that von Pettenkofer was wrong. Cholera is spread by the bacterium, and not by miasmas. So how did von Pettenkofer survive? There are a number of possibilities. Firstly, von Pettenkofer may have had high levels of stomach acid, which can protect against cholera when it is ingested in this way. Secondly, the cholera sample in the beaker may have died or degraded, perhaps through the presence of an antibacterial mould or bacteriophage. Thirdly, it's possible that von Pettenkofer may have been immune to cholera; either because he had had a very mild bout when he was younger, which he may not have even been aware of, or alternatively he may simply have been resistant.

The great Russian composer Peter Ilyich Tchaikovsky did something rather similar, but sadly he was not so lucky. Days after the premiere of his powerfully emotional Sixth Symphony, the Pathétique – which according to some accounts had been greeted with silence from its baffled audience – he drank a glass of cold, unboiled water in a public restaurant, which was an unthinkable thing to do during a cholera epidemic, and died as a result. The dangers of cholera and unboiled water were widely known, and so Tchaikovsky's dramatic, public act has been interpreted by many as a deliberate suicide.

How accurate is the presentation of medicine on TV and in films?

You probably won't be surprised to learn that the presentation of medicine, hospitals and surgery in film and television is full of inaccuracies. And yet many viewers do tend to believe what they see on the screen. A study in the year 2000 found that more than 50% of TV viewers trusted the medical information on TV, and about 25% of people described TV as one of their top three sources of medical information. The following are just a few examples of why this may be a mistake:

Chloroform – In any kind of TV drama, we all implicitly understand that one whiff of a chloroform-laced handkerchief will render any person silently and harmlessly unconscious within a second or two. In reality, the effects of chloroform are not nearly so predictable. Chloroform often takes some time to act, perhaps as long as a minute, and in that time the victim would be likely to struggle, shout, fight back, and generally make things difficult. It's also very easy to give too high a dose of chloroform, and this can cause vomiting, brain damage, and even death. One reason that chloroform is hardly ever used as an anaesthetic these days is that the margin between adequate anaesthesia and a potentially fatal dose is far too fine a line.

Trauma – In TV and films, it's quite normal for characters to get whacked with bottles, chairs, guns, and a whole range of other weapons, without suffering any real consequences. The whackee will usually fall over, and then get up again a bit later, and they're basically fine. In reality, just one blow to the head of this type can cause memory loss, concussion, brain damage, the loss of vision or hearing, and memory loss. When the victim regains consciousness, if they've been fortunate enough to avoid suffering a blood clot or slipping into a coma, they will certainly be extremely groggy and nauseous, and in no fit state to leap into a car and chase the bad guys into the next scene.

CPR – Cardiopulmonary resuscitation, or CPR, is that thing that people do in hospital dramas when they pump the patient's chest with their palms, and give them mouth-to-mouth (or, increasingly, ventilate the lungs using a bag or some other device). A recent study found that in TV dramas, more than two-thirds of CPR patients survived. In reality, the number is much lower, around 15%.

Heart attacks – This is another example of TV shorthand that we all instinctively understand: the fat man clutches his heart, gasps, and keels over and dies, and every viewer knows that this is the code for 'heart attack'. However, heart attacks rarely look like this in real life. They don't generally cause instant death, and often there is no sharp, sudden pain in the heart. This misconception has real, harmful consequences, as people are often slow to call an ambulance when they do have a heart attack, because they don't realise they are having one. The real symptoms of a heart attack are usually much milder than TV suggests. They include nausea, shortness of breath, sweating, dizziness, numbness in the arms, and pain in the chest, neck or shoulders, but usually not the sharp, stabbing kind we expect. Women in particular almost never experience the sharp chest pain so beloved of TV drama.

Defibrillators – In any hospital drama, no matter how bad the patient's condition, there is always one last resort. If he appears to have died, and the monitor has flatlined, you can always just shock him back to life, with the two magic electrical paddles, the defibrillators. In reality, paddles are rarely used these days, as sticky pads are applied instead. More importantly, defibrillators simply don't do the thing that drama producers seem to think they do. They don't shock the heart back into life. All they can do is correct a heart's rhythm. There are also no sparks in real life, and the patient's body doesn't leap into the air in a dramatic convulsion.

Guns – Most action heroes would be deaf by the third reel. Gunfire is extremely loud, and frequent exposure to it can cause permanent hearing loss. Handguns and automatic weapons fire at around 140 decibels, which is defined as being at the threshold of physical pain, and can cause instant hearing loss. Larger weapons such as bazookas and rocket-launchers are even louder. Guns can also damage your eyes. Shooting ranges provide safety glasses to protect the eyes from ejected shells and gunpowder dust, but this is never an issue on TV.

Another issue is that characters in films never fire guns properly. The recoil from even a small handgun can knock a man over if he's not properly braced, firing straight from the shoulder. In real life, police officers don't shoot one-handed, or side on (in fact, the majority of US police officers never fire a gun in the line of duty throughout their entire career). Thanks to dubious Hollywood gunplay, a common injury among US gang members is a broken thumb, caused by the recoil from mimicking the trend for sideways firing. As for firing automatic weapons with one arm like Rambo, it's not advisable, as it will very likely dislocate your shoulder.

Finally, being hit by a bullet does not send the victim flying backwards. Bullets are small, they don't weigh very much, and most of their energy is spent in travelling through your body, rather than pushing it backwards. Many people who have been shot report that they didn't feel a thing when the bullet hit them; the pain and shock tend to come later, when you actually see the wound and the blood.

What was surprising about the press conference which finally announced a proven link between smoking and lung cancer?

During the first half of the 20th century, the number of deaths from lung cancer began to grow rapidly. As contagious diseases such as tuberculosis and smallpox increasingly came under control, the deaths these diseases would have caused were being replaced by non-contagious causes of death: lung cancer, heart attacks, and strokes. Between 1905 and 1945, the rate of lung cancer in men increased twentyfold. In 1950, the number of deaths from lung cancer exceeded those from tuberculosis for the first time.

However, no one knew what was causing this increase in lung cancer. At the time, few people believed that smoking was harmful. There were restrictions on cigarettes being sold to children, but only because it was believed to stunt their growth. In 1950, Richard Doll and Austin Bradford Hill published the first major study to demonstrate that lung cancer was caused by smoking. As a result, Doll himself gave up smoking.

The 1950 study was a case control study, which compared one group of 709 people with lung cancer, with another group of 709 who were as similar to them as possible in every respect, except that they did not have lung cancer. The aim of a case control study such as this is to see if, all other things being equal, one factor leaps out as being distinctive to one group. In this case, it did. Of the men who had lung cancer, only 2 were non-smokers, whereas 27 of the men without lung cancer didn't smoke. This was a statistically significant finding, which would be confirmed by further studies in 1954 and 1957 by the same authors.

By 1954, the case was undeniable: smoking tobacco enormously increased the risk of lung cancer. The Conservative government's Minister of Health Iain Macleod held a press conference, in which he announced that lung cancer was caused by smoking. However, there was one extraordinary detail about this press conference – Macleod chain-smoked throughout it.

Why is it that so many murderers are doctors?

There do seem to be a disproportionate number of medical professionals among history's ranks of murderers, but it's not clear why this should be. It's possible that doctor-murderers are more newsworthy and memorable: when a doctor kills it is particularly shocking and frightening, because we invest so much trust in our doctors. Doctors are in a unique position to cause harm if they so choose, are often able to cover their tracks, and most people's knowledge of medicine is so rudimentary that the balance of power and knowledge is firmly in the doctor's favour.

It could be the case that doctors kill more often simply because it's easier for them to do so. Doctors have access to a wide range of poisons, and an expert knowledge of how they can be used. A doctor will know better than most people which poisons can be hidden in food or drink, how long it will be before they take effect, which other conditions the symptoms might resemble, and so on. A doctor is also in a unique position to administer such a poison, and less likely to be squeamish about the outcome. Many doctors also have access to surgical tools and other lethal weapons, as well as a particular knowledge of the human body, where it is weak, and how it can be damaged, numbed, put to sleep, cut open, and so on. But do any of these explanations account for the actions of the following men?

First of course, comes Dr Harold Shipman, one of the most prolific serial killers in history. He is known to have been responsible for an astonishing 215 murders, but the actual number is likely to be much higher. Most of Shipman's victims were his own elderly female patients. He would give them a lethal dose of diamorphine, sign their death certificates himself, and then forge their medical records to give the impression that they were already unwell. In the year 2000, Shipman was sentenced to 15 consecutive life sentences, and the Home Secretary ordered that he never be released. In 2004, he committed suicide, hanging himself in his cell. Amazingly, given the stories that follow, Shipman is the only doctor in British legal history to have been found guilty of murdering his patients.

Dr Edme Castaign was executed in France in 1824. He killed a wealthy patient named Hippolyte Ballet and his brother

Auguste using morphine, making him the first person to use the new drug to commit murder.

G.P. John Bodkin Adams was suspected of killing more than 160 of his patients in the 1950s, most of whom died in 'suspicious circumstances', and left him money in their wills. In 1957, Adams was charged with murdering wealthy patient Edith Morrell, with a lethal dose of barbiturates, heroin, and morphine. Even though he had arranged for her cremation to take place on the same day as her death, Adams was somehow acquitted.

Dr Buck Ruxton of Blackpool in 1935 cut up the bodies of his wife and his maid, covering the carpets and curtains with blood. When he was quizzed by the police, he said he'd cut his hand on a tin of peaches.

Dr Henry H. Holmes was America's first serial killer. He opened a hotel in Chicago for the 1893 World's Fair, which he built and designed himself. The upper two floors were a maze of dead ends, windowless rooms, stairways to nowhere, and other features designed to confuse the unwary. He would lure his victims into this labyrinth, including a number of his employees and girlfriends, and torture and kill them. Some of the rooms had gas lines, through which he would asphyxiate his victims. The victims' bodies would be dumped in a chute which led down to the basement, where Holmes would dissect them, or sell them to medical schools. In 1896, Holmes was hanged at Philadelphia County Prison. Over the course of his life, he is believed to have killed as many as 100 people, in cities all over the US.

In the 1930s, Dr Morris Bolber of Philadelphia carried out an audacious scheme, murdering thirty patients either by poisoning, or by whacking them round the head with a sandbag, so that he could collect their insurance money.

Dr Crippen murdered his wife with an overdose of hyoscine, and buried her body in the cellar (apart from her head, which is still missing). He then fled to Canada with his lover on an ocean liner. During the voyage, she disguised herself as a boy through-out. However, Crippen was recognised on the ship, and a series of morse code communications led to him being arrested on his arrival in Canada.

Finally, Dr Palmer of Rugeley had a reputation for extravagant living, and an eye for the ladies. He was also a compulsive gambler, but not a successful one. After one of Palmer's horseracing friends, John Parsons Cook, won a large sum of money at Shrewsbury, Palmer invited Cook to dinner to celebrate. After dinner, Cook became violently ill, and died two days later. At the post-mortem, Palmer was spotted trying to pocket Cook's stomach. He then tried to bribe various people involved with the coroner's office. Eventually, Palmer was charged with murder, after it was found that he had bought strychnine days before the murder.

CHAPTER SEVEN

THE HUMAN BODY

'The art of medicine consists in amusing the patient while nature cures the disease.'

Voltaire

What are hiccups for?

When we hiccup, what physically happens is that our diaphragm pulls down abruptly, drawing air into the lungs. However, within 35 milliseconds, the opening at the top of the air passage, which is called the glottis, slams shut, causing that strange 'hic' sound. Most bouts of hiccups end quickly, but some unfortunate people suffer for days or even weeks on end. The world record is held by Iowan Charles Osborne, who hiccupped continuously for an astonishing 69 years until his death in 1991 (which was not caused by hiccups). Despite his condition, Osborne managed to lead a relatively normal life, even somehow marrying, and conceiving eight children.

So what causes hiccups? There are many theories, and it seems that there may be a number of different causes. Some people say that hiccups are caused by laughing, gulping, drinking alcoholic or fizzy drinks, or swallowing too much air. Some blame spicy foods, or rich foods, or overeating. There are medical conditions which can cause hiccupping, including epilepsy, diabetes, tuberculosis, and bowel obstruction. One report suggests a man hiccupped for four days straight because a hair inside his ear was tickling the eardrum.

One persuasive theory holds that hiccups can be caused by the vagus and phrenic nerve systems. Persistent hiccups are often caused by nerve damage; for example, chemotherapy can cause hiccups, as can tumours found at the site of these nerves. One persistent hiccup sufferer was found to have a brain tumour; only after two-thirds of the tumour had been removed did the hiccups stop.

An interesting theory has recently emerged that attempts to explain why we have this useless, annoying reflex. Christian Straus from the Pitie-Salpetriere Hospital in Paris has argued that hiccups are a vestigial evolutionary relic, left over from a very distant ancestor species which had gills like a tadpole. Hiccupping is very similar to the way in which primitive air breathing creatures such as tadpoles gulp in air, while almost simultaneously closing their glottis to prevent water getting into the lungs. Ultrasound scans have revealed that babies start hiccupping in the womb after just twelve weeks, before any breathing functions have developed. It seems therefore that hiccupping may be an important stage in the baby's growth, preventing amniotic fluid from entering the lungs. If so, it would explain why humans still retain the ability to hiccup, even though as adults it serves no useful purpose.

Yes, yes, yes ... but how do you cure them?

I'm afraid there's even less consensus on this point. Everyone seems to have their own pet theory, but there is no strong evidence in favour of any one. Some of the more common suggestions include drinking water (often in some bizarre contorted position), drinking vinegar, blowing up a paper bag, receiving a fright, controlled breathing, sneezing, sucking, holding your breath, pulling your tongue, pressing your eyeballs, or eating a spoonful of sugar.

If the bout goes on for a dramatic length of time, there are also medical interventions that can be tried. Options include hypnosis, acupuncture, a catheter through the nose, or drugs, including sedatives and anti-spasmodics. However, these are only for seriously persistent cases, as effective treatment with sedatives usually requires a dose that will knock the patient unconscious.

How long does it take to digest chewing gum?

I'm sure we all remember being warned as children not to swallow our chewing gum because, as we all know, it stays in the stomach for seven years. In fact, this is an urban myth, which has presumably emerged from our observations that a) chewing gum isn't food, and so therefore shouldn't be swallowed, and b) chewing gum seems to have the strange property of being indestructible. It's easy to see how these two observations could merge into an assumption that chewing should not be swallowed *because* it is indestructible.

In fact, chewing gum doesn't usually remain in the stomach for any longer than other foodstuffs. The two functions of our digestive system, broadly speaking, are to digest those things we can use, and expel the rest. There are many things that the stomach struggles to break down, such as sweetcorn kernels, but it's not a problem, because these simply get expelled in our poo. Chewing gum is similarly indigestible, so it gets expelled in the same way.

Chewing gum generally consists of gum base, flavourings, sweeteners, and softeners. The last three can all be easily broken down and digested, which is why gum quickly loses its flavour, sweetness, and softness as we chew it. Gum base, however, is indigestible. Today, gum base is usually made of butyl rubber, which is also used in the manufacture of inner tubes for tyres. In the early 20th century, gum base was usually made of the natural resin from the chicle tree, which is native to Mexico. Mexicans have chewed chicle resin for centuries.

In fact, people have chewed all sorts of gums for thousands of years. Lumps of tree tar with tooth prints in them have been found in Finland dating back 5,000 years; the Finns are believed to have used these lumps to improve oral hygiene, much like chewing gum today, and the tar may also have had antiseptic properties. The ancient Greeks chewed gum, as did Native Americans. Today, the US military provides chewing gum to troops, as it helps to improve concentration and aids oral hygiene. The US army have also begun providing chewing gum laced with caffeine, to keep soldiers alert in the field.

Although chewing gum doesn't usually get stuck in the stomach, it is possible to create a blockage if you swallow enough of it, quickly enough, in combination with other indigestible items. A recent report in the Scientific American cited three cases of children who had swallowed large quantities of gum, along with coins, or sunflower seeds, and had consequently managed to create gooey, lumpy masses that did get stuck in their intestines, requiring medical treatment. So the answer is clear: chewing gum doesn't get stuck in the stomach. Except when it does.

What are the Crypts Of Lieberkühn?

It sounds like the title for a new Harry Potter adventure, but in fact the crypts of Lieberkühn are part of the human body. They are glands found in the lining of the small intestine and colon. In the context of anatomy, the word 'crypt' means a pit or a dead-end tube in an otherwise flat surface. The crypts of Lieberkühn produce enzymes, including sucrase and maltase, which break down food. The crypts were named after the 18th century German anatomist Johann Nathanael Lieberkühn, who is now best remembered for the medical specimens he produced by injecting wax into various organs and bodily cavities. The specimens Lieberkühn produced are regarded as masterpieces, and continued to be used for more than a century after his death.

In fact, many of the great names of medicine live on through the obscure parts of the body that bear their names. The following are, in my humble opinion, some of the more exotic and amusing examples:

- The fissure of Rolando is a fold in the cerebral cortex of the brain, named after Italian anatomist Luigi Rolando.
- The foramen of Winslow is a connection between two cavities of the abdomen, named after Danish anatomist Jacob Winslow.
- Adamson's fringe is a part of the hair follicle, where the active hair bulb meets the inactive hair shaft. It was discovered by dermatologist Horatio George Adamson.
- The ampulla of Vater is formed where the common bile duct meets the pancreatic duct. It was first described by a German anatomist named Abraham Vater.
- The sheath of Schwann is a microscopic membrane surrounding one of the cell types which make up a nerve fibre. It is named after German physiologist Theodor Schwann, whose many accomplishments include discovering the organic nature of yeast.
- The circle of Willis is a circle of arteries that supply blood to the brain, named after Thomas Willis, a pioneering researcher into the anatomy of the brain.
- Scarpa's triangle is a region on the upper thigh, where a number of major arteries and veins are found directly under the skin, making it an important site for surgery. It was named after Antonio Scarpa, a leading Italian

anatomist, whose head was removed and exhibited at the Institute of Anatomy after his death.

Do people go mad when there's a full moon?

For thousands of years, people have known that the moon's cycle affects our behaviour. The words 'lunacy' and 'lunatic' come from the name of the Roman goddess of the moon, Luna. In the 17th Century, England's Chief Justice Sir William Hale was in no doubt about the moon's power when he stated, 'The moon has great influence on all diseases of the brain, especially dementia.'

And the evidence continues to pile up. We have all heard anecdotal evidence: how Accident & Emergency departments are always busier during a full moon, how schoolchildren become even more unruly, and how violent crime goes up. But as well as anecdotes, there are also serious studies. In the year 2000 a Finnish research project found that people were more likely to commit suicide when there was a new moon. In 2007, a police force in Brighton examined the crime statistics and found evidence of a direct link between the full moon and increased violence on the streets.

And yet, despite all this evidence, the link between the moon and our behaviour is in fact a myth. There is no meaningful correlation between the cycles of the moon and any of the reported human behaviour. But how can that be, you may be wondering, given the compelling evidence already cited?

Well, there are a number of factors that can explain this discrepancy. Firstly, there is something called random variation: if you look at any set of statistics, some kind of pattern is likely to emerge. The question is whether or not this pattern is statistically significant and repeated. Although it's true that there are a few studies which do appear to show behaviour changing during specific parts of the moon's cycle, there are also many others showing no effect, or showing the opposite effect. For example, two studies published in the December 23, 2000 issue of the British Medical Journal looked at the rate of dog bites around the full moon: one study found that rates increased, the other study found that they decreased. Taken together, they suggest that there is actually no effect.

In 1986, scientists at the University of Saskatchewan carried out a meta-analysis, combining thirty-seven separate studies which had looked at the moon's effect on various aspects of human behaviour. Overall, the scientists concluded that many of the studies' reported findings were based on small, statistically insignificant samples, that the studies' findings frequently contradicted one another, and that overall there was no compelling evidence of any lunar effect.

Problems also arise through simple human error. Statistics is a complicated business, and many studies are flawed. The same meta-analysis mentioned above looked at twenty-three studies which appeared to show a correlation between human craziness and the moon's cycle, but found that nearly half of those studies contained at least one statistical error. Other meta-analyses have since confirmed these findings: there is no overall evidence of any correlation, and those studies that do appear to show a correlation are often methodologically flawed.

This problem is exacerbated by the media, which only ever reports those studies which appear to confirm the crazy moon theory. This is to some extent understandable: when a study finds that there is, say, no link between homicide rates and full moons, that's not an amusing or newsworthy story. When, however, a study comes out a week later that does find a link, that makes a great story, it doesn't do any real harm, and editors have got pages to fill.

Another factor is a tendency known as 'confirmation bias', which essentially means that people are more likely to notice things which fit into their existing beliefs or hypotheses.

Therefore, if you work in an A&E department, for example, and you've heard that it gets busy when there's a full moon, you may well pay particular attention on those nights when it does happen to be busy and there is a full moon. Perhaps you have a coffee with your colleagues when it's over, and swap stories of other crazy full moon nights. However, there will also be busy nights when there's no full moon, and quiet nights when there is, but people are less likely to notice or remember these.

The final objection is that it's not at all clear how a full moon could conceivably have any meaningful effect on human behaviour. Those who believe in the 'lunar effect' tend to mention the tides. They argue that 80% of the earth's surface is water, just as 80% of the human body is made up of water, and that therefore our bodies might, like the tides, be affected by the moon's gravity. However, there are a number of problems with argument. Firstly, when the moon is full, it just means it is more visible than during other phases, it doesn't have any more gravity. A new moon is just as close to the earth as a full moon, so it ought to have the same gravitational effect. The second problem is that the moon's gravitational effect on the tides, and on us, is tiny; it is estimated that a mother holding her baby is exerting around twelve million times as much gravitational force on the child as the moon, simply because she is closer to it. One astronomer has estimated that a mosquito landing on your arm exerts more gravitational force on you than the moon.

Ah, but we already know that the moon can affect people, because we know it affects the menstrual cycle!?

I'm afraid, hypothetical questioner, that the evidence suggests that this theory too, though widely held, is actually false. The apparent reason for the development of this myth is that the lunar cycle, which is the length of time between two full moons, is 29.53 days, which is similar to the average length of the menstrual cycle. However, although the average menstrual cycle is about 28–29 days, the length of women's cycles varies enormously: women can have regular cycles as short as 21 days, or as long as 36. Longer or shorter menstrual cycles will, by definition, be out of sync with a 29.53 day lunar cycle. However, cycles of around 26–30 days might appear to be in sync just often enough to reinforce this myth.

The menstrual and lunar cycles may also appear to be synchronised because there are a number of different phases of the menstrual cycle, and a number of different phases of the moon, many of which generally last for a few days rather than a single day, making it easier to link any two of them together. For example, one study in 1980 found that 40% of women in a random sample began their periods during the 'light half cycle', meaning the two weeks around the full moon. However, two weeks is a very broad period of time. If women's cycles were not synchronised with the moon in any way, we would expect around 40% of them to begin in any given two week period. In other words, 40% is exactly the result we'd expect if the moon had no effect on women's cycles.

The other related belief is that women synchronise their menstrual cycles with one another, when living or working in close proximity, and again I'm afraid the evidence suggests that this too is false. For a start, it wouldn't make any sense if menstrual cycles were also synchronised with the moon, as they can't be synchronised with both. Obviously, it's my contention that menstrual cycles are not synchronised with the moon, but the point is that people who believe the one tend to also believe the other.

The menstrual synchrony theory received a major boost in 1971 when Martha McClintock published a paper in Nature magazine which found that women in her dorm at Wellesley College, Massachusetts tended to synchronise their periods.

However, the study only suggested that the women's cycles were, on average, within about 5 days of one another. This may sound compelling at first, but it's pretty much what we would expect from random variation in such a small sample. If the average cycle is 28 days, then two women's cycles could not be more than 14 days apart; so this is the maximum difference. The minimum difference is obviously zero. We would therefore expect an average of about 7 days' difference, with just as many incidences of cycles being fewer than 7 days apart as there are incidences where the gap is greater than 7 days. In this context, in a small sample, an average of 5 days is not particularly surprising.

There were also a number of methodological problems with McClintock's study. Firstly, the women reported their dates themselves, retrospectively, which creates the possibility of confirmation bias. Secondly, which cluster of dates do you decide to use? In any small sample of random data such as this, it's fairly easy to pick a point where there is an apparent cluster. Further studies have attempted to replicate these results, but have found no evidence of menstrual synchrony.

Where in the human body is the soul?

The French philosopher René Descartes was fascinated by a tiny gland called the pineal gland, which is found deep in the centre of the brain. This tiny teardrop-shaped gland seemed to serve no purpose, and appeared to be the only part of the brain which consisted of one single unit, rather than two matching pairs or halves. Since our senses generally work in pairs – we have two eyes, two ears, two hands, and so on – Descartes' theory was that there must be a central hub in the brain, where information collected in this dual fashion was brought together and assimilated, to be presented as a single sensory experience to that part of us which we call the mind or the soul. Therefore, Descartes concluded, the pineal gland must be where our soul is located, and where all our thoughts and ideas are formed.

Today we know a little more about how the pineal gland works, along with the rest of the brain. We now know that the pineal gland does have a number of functions. It produces a hormone called melatonin, which regulates puberty. It also regulates our circadian rhythm, which is our body's biological clock, which determines which part of the day we feel most awake, most drowsy, most coordinated, and so on.

However, people of a mystical or spiritual bent have a number of more exotic theories about this mysterious part of the body. It is sometimes described as our 'third eye', because it seems to have some evolutionary link with the photoreceptive parietal (or third) eye which is found in some other animal species, including lizards, frogs, and lampreys. There are claims that the pineal gland can secrete a natural psychedelic drug called DMT, which is somehow connected to near-death experiences, such as out-of-body sensations, visions of a light at the end of a tunnel, encounters with angels, and so on. The pineal gland also seems to affect our responses to certain drugs, such as cocaine and anti-depressants.

The pineal gland does not seem to be the site of the soul, then, but it remains mysterious and poorly understood. In 2007, an experiment in China using MRI scans found that the area surrounding the pineal gland becomes more active during a religious meditation technique called Chinese original quiet sitting, in which practitioners silently recite religious mantras. Although this was a very small-scale study, it has added to gland's exotic mystique.

Do taxi drivers have bigger brains?

To drive a London black cab, taxi drivers have to pass an extremely demanding course called 'The Knowledge', which requires them to develop an intimate familiarity with every street and place of interest in the city. It can take three years to learn, and more than three-quarters of those who start the course drop out.

In the year 2000, a study by Eleanor Maguire at University College London found that London cabbies tend to have one region of the brain which is significantly enlarged, compared with the rest of us. At this point, you might be wondering what this means. Does it indicate that only people with a particular type of brain are capable of learning 'The Knowledge'? Or, alternatively, does it suggest that driving a taxi enlarges a particular section of the brain? In fact, the study in question does attempt to distinguish between these two possibilities. The research team found that the size of the drivers' posterior hippocampus region was positively correlated with the length of the time they had been driving a taxi – in other words, the longer they'd been a taxi driver, the bigger this region would be – which would seem to indicate that driving a cab does make your brain expand.

However, the researchers also found that there was a trade-off. While taxi drivers did develop a larger posterior hippocampus, which suggests that this part of the brain is used in navigation or memory, another section of the brain called the anterior hippocampus would shrink by an equivalent amount to compensate, so overall their brains were no bigger than average. This may even mean that taxi drivers lose some of their other faculties, in the course of developing their navigational expertise.

What percentage of the brain do we use?

It is a well known fact that we only use 10% of our brain, and that therefore we have enormous capacity to improve our intellectual abilities, and perhaps harness mysterious, spiritual powers. Sadly however, this is another of those well known facts which turns out to be false. The very specific nature of this claim – it is almost always 10%, rather than, say, 'a small proportion' – adds to the impression that it is based on a particular scientific study, but in fact there is no such study, and this 10% myth seems to simply be a misconception that has somehow taken hold.

The idea that we only use 10% of our brain seems to have developed during the late 19th century. It has been attributed to the pioneering psychologist William James, who stated in his book 'The Energies Of Men' that, 'We are making use of only a small part of our possible mental and physical resources.' The anthropologist Margaret Mead reportedly said something similar, and Albert Einstein made a quip along the same lines, in regard to his own intelligence. One way or another, the idea took hold that this was a proven, scientific fact.

However, it's hard to see how it could actually be true in any meaningful way. If 90% of our brain was really not being used, then people could lose 90% of their brain in an accident, and suffer no ill effects. In reality, a small amount of damage to any area of the brain will usually reduce a person's abilities to a dramatic degree. If 90% of our brain were removed, we would be left with about 140 grams of brain, which is about as much as a sheep. Sheep are cute, but they're not very clever.

Another objection is that most parts of our body tend to wither if they are not used. When people break a leg, and have to spend weeks resting with their leg in a plaster cast, the limb becomes visibly weakened through underuse. If 90% of our brain was never used, we would expect adults to have shrunken, withered brains, but they don't.

Furthermore, if 90% of our brain were never used, it seems likely that we would probably have evolved to have smaller, less demanding brains, and smaller skulls, to make childbirth less painful and dangerous. The human brain brains makes up only around 3% of our body weight, but it uses up an enormous 20% of our energy and resources. If 90% of the brain were made of useless material, which was nevertheless costing us all this energy, it seems likely that over time we would have evolved a more efficient system.

Although the 10% 'fact' is false, it is possible to imagine how the misconception may have arisen. One possible explanation is the fact that only 10% of our brain consists of neurons, while the other 90% is made up of glial cells, which encapsulate and support the neurons. In the past, glial cells were thought to have only a very limited function, but recent studies have shown that, like neurons, glial cells have chemical synapses, and release neurotransmitters. However, people may have interpreted this ratio and drawn the conclusion that the only important part of the brain is the neurons, which make up 10% of the brain's cells. Alternatively, the misconception may have arisen in reference to the number of neurons which are firing at any one time, which might theoretically be around 10%, although it's not immediately obvious how this could be measured. However it arose, it seems certain that the theory is false. Brain imaging scans of various types have shown that we use every part of our brain, and at any given time most sections of the brain are active, even while we are asleep.

Are women's breasts getting bigger?

They are indeed. In 1983, the best-selling bra sizes in the US were 34B and 36B; today the best-seller is 36DD, which is a significant increase. In the UK, Marks and Spencer bras used to go up to a size G, but in 2007 the company announced that it was introducing sizes GG, H, HH and J, to accommodate the larger busts of modern women. So what's going on here?

The obvious answer is that we are simply getting bigger in general. Since 2007, the rate of obesity in American women has grown by 2.1% to 35.3%. Today's women are also better nourished than in the past, which means they are taller, with stronger frames. The growth in breast implants may also be a factor, although probably not a significant one. In 2007, just over 300,000 breast augmentation procedures took place in the US, which may sound like a lot, but it isn't a very big number in a population of 155 million women. Another factor is increasing use of the contraceptive pill, as hormone levels are a significant factor in breast size.

Some people also credit improved education about properly-fitting bras. According to various surveys, anywhere between 70–90% of women wear the wrong size bra, and some experts suggest that the very system of bra measurement is so flawed as to be useless. In November 2005 Oprah Winfrey devoted an entire show to bras and how to find the correct fit, and this TV event encouraged vast numbers of women to have a proper fitting carried out. In many cases, this meant changing to a smaller band size, but a larger cup size.

Although breasts do seem to be getting bigger, there is also something intriguing called 'cup inflation' going on. It is similar to the reverse trend in fashion retailing, whereby the proportions of each dress size are becoming bigger, which means that larger and larger women can increasingly squeeze into dress size numbers which were previously cut to smaller specifications, and consequently feel good about themselves, because they are still, say, a size 12.

This is one reason for the enduring fascination with Marilyn Monroe's dress size. It is widely claimed that Monroe was a size 16, which has led to a lot of debate about whether models today are too skinny, or standards of beauty have changed, and so on, but in fact, even if Monroe was a size 16 in the 1950s, that doesn't mean she would be a size 16 today. Recently, a British fashion writer was given the chance to try on some of Monroe's dresses, and found that they were equivalent to a UK size 8 or 10 today, which is roughly a 4 or 6 in the US.

Cup inflation of bra sizes, on the other hand, is moving in the opposite direction. Increasingly, manufacturers are labelling bra sizes bigger than in the past, so that customers will feel reassuringly buxom and sexy. A woman who was a 36C a few years ago might well now be a 36DD, without any actual change in the size of her breasts.

Furthermore, it's not just women's breasts that are getting bigger; men's are too. The problem of man-boobs, or gynecomastia to give the condition its proper name, is a growing one. It is thought to be caused by hormone imbalances, and by excess body fat. Weight loss can help the problem, but many patients also consider surgery.

Is the tongue the strongest muscle in the body?

This is another of those well known facts about the body that turns out to have no credible basis. It's hard to imagine any definition of strength by which the tongue would be considered the strongest muscle in the body. The most sensible definition of a hypothetical 'strongest muscle' would seem to be the one which can apply the greatest direct, measurable force on some external object, in which case the winner is the masseter muscle. The masseter is the jaw muscle, and there are two of them, on either side of the jaw. The masseter has a significant advantage over other muscles in that it uses the bones of the jaw as powerful levers. As a result, jaw muscles have been recorded delivering a bite force of 975 lbs for two seconds, according to the Guinness Book of World Records.

Alternatively, if we consider muscles without the mechanical advantage of levers of this type, then the strongest muscles are simply the biggest ones, as individual muscle fibres are generally all of similar strength. By this measure, the biggest muscles are either (depending on the individual) the quadriceps, which is the front thigh muscle, or the gluteus maximus, which is basically the bum.

However, there are a number of other ways in which we might define muscle strength. Pound for pound, shorter muscles tend to be stronger than longer ones. Relative to its weight, the myometrial layer of the uterus is the strongest muscle in the body, as the entire uterus weighs only about 2.5 lbs, but exerts a force during childbirth of as much as 100 lbs. To look at the question in another way, the heart does more work over the course of a lifetime than any other muscle. Most muscles quickly get tired when given work to do, but the heart beats continuously. In short, there are quite a few different ways to work out what we might consider to be the strongest muscle in the body, but the tongue doesn't seem to be the right answer under any circumstances.

CHAPTER EIGHT

RUDE BITS

'Always laugh when you can. It is cheap medicine.'
Lord Byron

What is 'sounding'?

Before the development of antibiotics in the 20th century, there was no effective treatment for gonorrhea, which is a common sexually transmitted disease that can cause sterility. Male sufferers find it painful to urinate, and may produce a thick yellow discharge, which is known as 'gleet'. Infected females may have similar symptoms, but many are often symptomless. However, if the infection is allowed to spread, it can cause severe complications.

Gonorrhea can cause scars and obstructions to grow inside the urethra, which is the tube that leads from the bladder to the genital opening; it is the tube through which we urinate. In men, it runs through the length of the penis. In women, it emerges just above the vaginal opening. If the urethra is blocked, it can make urination painful or even impossible.

'Sounding' is a method of treating these obstructions. The doctor inserts a long metal probe into the opening of the urethra, and uses this device, which is called a 'sound', to stretch and widen the tube, clearing any blockage. It is said to be an extremely unpleasant experience for the patient, and men have been known to faint.

Nowadays, gonorrhea can be treated with antibiotics, so you would think that sounding would have died out, but surprisingly it lives on. The reason for this is that some people find it rather fun. Fetish websites sell sets of sounds, of varying shapes and sizes, along with guides as to their use. One apparently pleasurable practice is to insert a curved sound down the urethra, deep into the bladder, to stimulate the male prostate. There are even piercings which incorporate sounds. One of these, which is known as the Prince's Wand, consists of a urethral sound held in place by a Prince Albert piercing, which is a thick ring which is inserted into the glans of the penis. As if life wasn't painful enough...

Did Christopher Columbus bring syphilis to Europe from the New World?

This question continues to be the subject of heated and enduring debate. The theory is that when Christopher Columbus returned from his first trip to the Americas in 1493, some members of his expedition brought back syphilis, which they had contracted in the New World. Within two years, Europe had its first epidemic of syphilis, which would go on to plague the continent for another 400 years.

This is by no means a new theory. For centuries, people have believed that Columbus and his expedition brought syphilis to Europe, and there is considerable evidence to support the idea. The first European syphilis epidemic broke out in 1495 among French troops, who are believed to have included in their number many of Columbus's mercenaries. There is also archaeological evidence, as the bones of syphilis victims have distinctive lesions and pitted skulls. Lots of South American, pre-Columbian remains have been found with these markings, which suggests that syphilis was certainly present in South America before Columbus's arrival. A recent study looked at the molecular structure of 26 strains of the bacterium Treponema pallidum, and found that those strains which cause venereal syphilis were the ones which had originated most recently, and were closely related to similar strains from South America.

However, the theory also has its critics, as there is some evidence that syphilis was already present in Europe before Columbus's return, but cases of it may often have been attributed to leprosy. As far back as the 4th century BC, Hippocrates described a disease which some believe may have been syphilis. 14th century skeleton fragments from England and Italy have been found which show possible signs of syphilis, although this is disputed. In response to the recent molecular study mentioned in the previous paragraph, a number of anthropologists have argued that the strains of Treponema pallidum exhibit fewer similarities than the study claims, and that therefore no firm conclusions can be drawn.

The issue is further complicated by the fact that the diseases Columbus took with him from Europe to the New World had a far more devastating effect than syphilis. The New World natives had no resistance or immunity to the germs introduced by the Conquistadores. The first epidemic may have been swine flu, transmitted by pigs from Columbus's ships. This was followed by epidemics of smallpox on Hispaniola, Puerto Rico, and Cuba. When the Spanish adventurer Cortés attacked the main Aztec city of Tenochitlán (today's Mexico City), he took just 300 troops. Three months later, the city had fallen, and more than half of the city's population of 300,000 had died, most of them from smallpox. More waves of disease followed, as Amerindian populations were decimated by measles, typhus, smallpox, and influenza. Many native populations were brought to the brink of extinction.

As a result, Columbus's arrival at Hispaniola has been described as 'the most disastrous event in the history of human health'.

Why were obstetrical forceps kept secret for over a century?

Giving birth in the 16th century was an extraordinarily painful and dangerous business. If there were any complications, the odds of mother or baby surviving were slim. Towards the end of the century, a pair of French brothers began to make their name in London. They were the sons of Huguenot surgeon William Chamberlen, who had fled France in 1576. Bizarrely, both sons had been named Peter, so they were known as Peter the Elder and Peter the Younger respectively.

The family business was surgery, and obstetrics in particular, even though male midwives were rare at this time. Peter the Elder became the surgeon and obstetrician to Queen Anne of Denmark, in London, and as the brothers' reputation grew, they became celebrated and wealthy, with Peter the Elder attending a number of senior royal births.

The reason for their success was that they had a tremendous secret, which they publicly alluded to. They hinted that they had certain techniques, or perhaps instruments, that allowed them to safely deliver babies in circumstances that other midwives would find impossible. Of course, if this were true, it could potentially be of enormous public benefit, and save many lives, but there was no hope of patenting such an instrument or technique, and so the Chamberlens felt that it was in their commercial interest to keep it a secret.

To protect their secret, they went to extraordinary lengths. They would bring the device into the birthing room in an enormous ornate box, which took two men to carry. The idea was clearly to give the impression that the box contained some huge, complicated machinery. The expectant mother would be blindfold, to prevent her from seeing the device, and no observers were allowed in the room. Instead, they would listen on tenterhooks outside the door, while ringing bells and strange mechanical noises hinted at magical technologies.

Of course, this was all smoke and mirrors. The device was simply the obstetrical forceps, which is a curved, two-bladed instrument which is still used today in difficult births. It is believed that Peter the Elder invented the device. The Chamberlen family business continued well into the 18th century, as Peter the Younger's son, grandson, and great-grandson all carried on the tradition, and no outsider managed to find out the secret of the Chamberlen family's remarkable success in delivering difficult births. At various points, the family even tried to sell the device, and eventually Peter's grandson Hugh is believed to have done so, selling the forceps in the Netherlands to a select group of doctors. However, the secret didn't come into the public domain until the 1720s, perhaps more than 120 years after its invention.

In 1650, the rate of maternal death in childbirth was around 160 for every 10,000 births. Thanks to the widespread adoption of obstretrical forceps, by 1850 this had dropped to just 55 deaths per 10,000 births. This means that for more than a century, almost three times as many women died in childbirth as was necessary. In 1813, the Chamberlen family forceps were finally found, hidden under a trap door in a loft at the family's grand country home in Essex.

Did a busty girl inspire the invention of the stethoscope?

For centuries, doctors would listen to a patient's heartbeat or lungs by simply placing their ear against the patient's naked chest. There are even reports of weary doctors laying their head on a soft, warm bosom and accidentally nodding off to sleep. However, some doctors found the technique ineffective, and somewhat embarrassing, and one of them was the French physician René Laennec.

One day in 1816, Laennec found himself faced with a patient whose generous embonpoint made such an inspection unfeasible. Laennec had an idea, perhaps aided by the fact that he played the flute: 'I happened to recall a well known acoustical phenomenon: if one places one's ear at the end of a beam, one can hear very distinctly a pin dropped on to the other end.' Thus inspired, he rolled up a few sheets of paper, fashioning a rudimentary stethoscope, through which he could listen to the girl's heartbeat from a safe distance, without having to go near any boobs.

To his surprise, Laennec found that his device was not just adequate, but clearly superior. He could hear the heartbeat far more clearly than he had ever managed by just laying his ear against the patient's chest. Fired with success, he set about inventing a proper medical tool. The first stethoscope he designed was essentially a long wooden cylinder, but he then went on to refine and develop this concept. Using his new device, he was able to observe and classify the different sounds produced by the heart and lungs to a much greater degree of detail and clarity than had ever been done before. Many of the terms used by doctors today to refer to the various sounds produced by the heart and lungs were first recorded and defined by Laennec.

What was unusual about Miss Jenny's launderette in Georgian London?

The answer is that it was a condom launderette, which was located on St Martin's Lane near Charing Cross. At Miss Jenny's, ardent young men could have their condoms washed so that they could be reused - in those days condoms were made of linen or animal intestine. Miss Jenny's customers could also buy second-hand condoms, which had already been washed and hung up to dry.

Modern condoms are usually made from latex, which wasn't invented until 1920, but for centuries people had used more rudimentary condoms made from other materials. Condom use may even go as far back as the Roman Empire, although the evidence for this is sketchy. The first unambiguous reference to condoms was made in 1564 in Gabriele Falloppio's 'De Morbo Gallico', which means 'The French Disease' – in other words: syphilis. Falloppio claimed to have invented a condom made of linen, to be tied onto the penis using a ribbon. Around this time, condoms were also being made from the intestines and bladders of animals such as sheep and goats. Archaeologists have even found surviving examples of condoms made from animal intestine dating back to the 1640s, at Dudley Castle in the West Midlands.

As the deadly syphilis epidemic spread throughout Europe, condom use became increasingly widespread. Casanova made a diary entry in 1753 describing his practice of buying condoms by the dozen. To test their quality, he would blow them up before use, to make sure there were no holes. James Boswell and the Marquis de Sade also made references to condoms, and prophylactics can be seen in the background of a number of contemporary paintings, hanging up to dry, ready for their next sortie.

To make a sheep-gut condom, the condom-maker would soak a sheep's intestine in water for a number of hours, before turning it inside out, and leaving it to soften in a mild alkaline solution. Once the intestine was adequately softened, the condom-maker would scrape away the mucous membrane, leaving the peritoneal and muscular coats, which were then treated with the vapour of burning sulphur. After this, the membrane would be washed, inflated, dried, and cut to a length of around seven or eight inches. Finally, the open end would have a ribbon attached, which was used to tie the condom onto the base of the penis. Generally, users would soak the condom in water before use, to make it more supple.

By the late 18th century, Londoners had a range of condom shops to choose from, and in this raucous time, before the prudish age of Victoria, these businesses would openly market their wares. One of Miss Jenny's more upmarket rivals was a Mrs Phillips, who ran a comprehensive sex shop in what is now Covent Garden, selling condoms, sex books, flagellation machines, and 'widow's comforters'. Mrs Phillips' advertising leaflets boasted of her 35 years' experience in making and selling her 'implements of safety', which even came with ribbons in regimental colours. Her handbill closed with a short verse:

> *To guard yourself from shame or fear*
> *Votaries to Venus, hasten here;*
> *None in my wares e're found a flaw,*
> *Self-preservation's nature's law.*

What is a UBI?

There can't be many of us who haven't wondered at some point quite what was written on our illegible prescription or hospital chart. Doctors are educated people, skilled with their hands; so why can't they write clearly? Well, one reason might be that they don't want us to know what they're writing, as it seems that sometimes, the joke is on us.

For example, if you ever do see the letters 'UBI' on your medical records, you might like to know that your doctor clearly believes that you are the suffering from an Unexplained Beer Injury. Doctors and nurses have a whole range of amusing acronyms and slang terms with which to express their true opinions about our often self-inflicted ailments, although more open access to patient notes nowadays means that these terms are more likely to be used in conversation and private notes than on a patient's formal records. Here are a selection:

Plumbum Oscillans – Latin for 'swinging the lead', with the implication that the patient is trying his luck for a sick note

DBI – Dirtbag Index, calculating by multiplying the patient's tattoos by the number of missing teeth

CTD – Circling The Drain

GPO – Good for Parts Only

Departure Lounge – Geriatric ward

LOBNH – Lights On But Nobody Home

Oligoneuronal – A fancy term meaning 'thick'

GOK – God Only Knows

BTSOOM – Beats The Shit Out Of Me

PAFO – Pissed And Fell Over

HTK – Higher Than a Kite

TTFO – Told To Fuck Off

FOS – Full Of Shit, a patient who is not considered to be entirely honest

FLK – Funny Looking Kid

AGMI – Ain't Gonna Make It

Code Yellow – A bladder control emergency

Code Brown – Must I spell it out?

DFKDFC – Don't Fucking Know, Don't Fucking Care

What was 'The Fruits of Philosophy'?

This was a scandalous book, the Lady Chatterley's Lover of its day. It was written by Charles Knowlton, who was a fairly unremarkable doctor in Ashfield, Massachusetts, except that he had literary ambitions. In 1832 he wrote a small pamphlet called 'The Fruits of Philosophy, or the Private Companion of Young Married People' which he showed to a few of his patients. Unlike Lady Chatterley's Lover, this was not a novel. It was a non-fiction guide to conception, with advice concerning infertility and impotence. However, it also contained tips for birth control, including a method for douching the vagina after intercourse.

Then as now, there were some people who felt that the idea of birth control was sinful. A campaign was begun by the town's minister, Mason Grosvenor, and Knowlton was prosecuted and fined. The book was then reprinted in Boston, and Knowlton was punished again, and this time he was given three months' imprisonment and hard labour. After his release, Knowlton continued with his medical practice, and developed an excellent reputation as a reliable and conscientious physician. He died of heart failure in 1850.

Twenty-seven years after Knowlton's death, the book caused a fresh sensation in London. It had been published by the provocative freethinkers Charles Bradlaugh and Annie Besant, who were consequently prosecuted under the Obscene Publications Act of 1857. The Solicitor-General was outraged by the book:

'I say this is a dirty, filthy book, and the test of it is that no human being would allow that book to lie on his table; no decently educated English husband would allow even his wife to have it. The object of it is to enable persons to have sexual intercourse, and not to have that which in the order of Providence is the natural result of that sexual intercourse.'

Bradlaugh and Besant were both given six months in prison, along with a £200 fine, although these were quashed on appeal thanks to a technicality. The trial however had the effect of turning the book into an overnight bestseller, selling more than 150,000 copies. The book's resulting fame is believed to have had a major, positive impact on the use of contraception on both sides of the Atlantic, and it is even credited by some with slowing Britain's population growth.

What was 'The French Pox'?

The answer is syphilis, or at least it was if you were English. As syphilis epidemics spread throughout Europe after 1495, the disease became known by various different names across the continent, and the origin of these names followed a clear pattern: each nation blamed syphilis on its worst enemy. In France, it was *Le mal de Naples*, while the Italians called it The Spanish Disease. The Japanese called it The Chinese Disease, while most of China blamed the Cantonese, those people from the province of Guangzhou.

In fact, syphilis is just one example of a broader linguistic trend. In England, the French were seen as being crude and sexually permissive, and so condoms became known as 'French letters', genital herpes was known as the 'French disease', and an open-mouthed kiss was a 'French kiss'. Because of this saucy reputation, 'French postcards' and 'French novels' meant pornography, and you would apologise for swearing by asking someone to 'Pardon my French'.

In France the same thing happened, only the other way round. On that side of the channel, *filer à l'anglaise* means to leave without permission, or without saying goodbye, literally 'English leave'. In England, the exact same concept is called 'French leave'. Our French letter is their 'capote anglaise', while French kissing becomes 'le baiser anglais'. *Plus ça change…*

What was the public's great fear, following the discovery of X-rays?

X-rays were discovered by accident in 1895, by a German physics professor named Wilhelm Röntgen who was experimenting with electrons in vacuum tubes. Waving his hand between the covered tube and a screen, he saw an image of the bones of his hand projected onto the screen. He then took an X-ray picture of his wife's hand, and this astonishing image caused an international sensation.

X-rays are a form of radiation that pass through different parts of the body in different ways, depending on the density of the body parts. Because of this, X-rays can produce images of the internal structure of the human body, and these images can be captured on photographic plates. The discovery of X-rays was a huge breakthrough, as it meant that for the first time doctors would be able to look inside the human body without cutting it open. Within weeks of Röntgen's find, doctors were taking pictures of broken bones, gallstones, lesions, and all sorts of objects lodged inside people's bodies, including bullets and needles.

Today we know that, while they can be incredibly useful, X-rays can also increase the risk of cancer, and therefore need to be used sparingly. In 1895, however, no one knew any of this. Instead the public's chief fear, it seems, was that X-rays could be used to look through people's clothes, to ogle their naked bodies. A company in London began marketing 'X-ray proof underwear', and is said to have made a small fortune. The New Jersey State Legislature even brought in a law forbidding the use of X-rays in opera glasses, on the grounds of protecting public modesty!

Interestingly, while these fears may now seem rather quaint and daft, today we face a fairly similar issue. Controversial new scanners have been introduced at airports, which do clearly show the naked bodies of passengers, including their genitalia. Airport staff have already been caught misusing the scanners – a male member of staff at Heathrow Airport was alleged to have used the scanner to take a photograph of an unwitting female colleague, who then complained of sexual harassment. At Manchester airport, two women were barred from boarding their flight after refusing to walk through the scanner. At Miami Airport, a fight broke out after a male member of staff was taunted about the size of his penis, as revealed by the scanner. Perhaps in the future these scanners will seem as harmless and trivial as X-rays, but at present this seems unlikely.

Who was Patient Zero?

In 1987, a gay investigative journalist named Randy Shilts released a book dealing with the early years of the AIDS epidemic, called 'And The Band Played On'. The book became a highly controversial bestseller, particularly for its most sensational claim: that one highly promiscuous gay man, the so-called 'Patient Zero', had played a major role in bringing AIDS from Africa to the US, and then spreading it from city to city. Patient Zero was named in the book as Gaëtan Dugas, a Canadian flight attendant, who had died of AIDS-related illnesses in 1984.

Most of the book's marketing and coverage focused on the Patient Zero angle. Time Magazine opened its review with the headline, 'The Appalling Saga of Patient Zero'. The National Review called Dugas, 'The Columbus of AIDS'. California magazine promoted its serialisation of the book with an advert showing a picture of Dugas over text which ran, 'The AIDS epidemic in America wasn't spread by a virus. It was spread by a single man…. A Canadian flight attendant named Gaetan Dugas…'

According to the book, Dugas was blond, handsome, charming, and extremely promiscuous. He would travel the world, picking up men everywhere he went. Arriving at a gay bar, he would survey the room, announcing with satisfaction, 'I'm the prettiest one.' After developing Kaposi's Sarcoma, which is a form of AIDS-related skin cancer, he was warned by doctors that he might infect others, but he carried on regardless, refusing to stop having unprotected sex. Sometimes after an encounter, he would gesture towards the purple lesions on his chest, telling his potentially doomed partner, 'Gay cancer. I have it, maybe you'll get it too.'

AIDS was a terrifying and emotive subject in 1987, and Shilts' book had the effect of personalising the disease, and allocating blame onto one man, and also implicitly onto a whole community. The account given of Dugas reinforced existing prejudices and stereotypes that already perceived gay people as promiscuous and irresponsible. AIDS now had a face, a Typhoid Mary for the 1980s, confirming the view that this was a disease spread and ultimately caused by the gay community.

However, this account of Dugas was not just unreasonable in the way it sought to apportion blame, it was also factually incorrect. Shilts' account was based on an earlier study, which had looked for links between a number of the first AIDS sufferers, and found that 40 of them were directly or indirectly linked to Dugas. No one was named in the study, and Dugas was referred to only as 'Patient O', with the 'O' standing for 'Out of California'. This was misunderstood as Patient Zero, a misunderstanding which implied, incorrectly, that Dugas had been the first person to bring the disease to the US. Furthermore, the suggestion that it was Dugas who had infected these men with HIV was also unlikely to be true, as AIDS can take years to show symptoms, whereas most of these cases were diagnosed within months of the relevant encounters.

A more recent study in 2007, which looked at HIV gene sequences in the blood samples of early AIDS sufferers, found that Dugas could not have been the first person to bring AIDS to America, as it had already arrived from Haiti, possibly as early as 1969. AIDS had already been circulating in the US for around twelve years before it was first identified in 1981.

What was Antony van Leeuwenhoek the first man to see?

Antony van Leeuwenhoek was a Dutch tradesman and scientist, born in 1632, who today is generally regarded as the father of microbiology. His career would have been an extraordinarily successful one for any professional scientist, so it is even more remarkable that van Leeuwenhoek was only an enthusiastic amateur. He had no university education, and no scientific training. His family were not wealthy, so he earned his living through a range of different, mundane jobs. And yet, through his skill and intellectual curiosity, he made some of the most astonishing discoveries in the history of science.

Perhaps inspired by the polymath Robert Hooke, van Leeuwenhoek began constructing his own simple microscopes sometime before 1668, and using them to make observations of anything and everything that took his fancy. Over the course of his life, he is known to have made over 500 different microscopes, although fewer than ten of them have survived to the present day. Van Leeuwenhoek's microscopes were not the compound microscopes used today, which use multiple lenses; instead they were more like high-powered magnifying glasses. However, van Leeuwenhoek's skill at moulding glass, grinding lenses, and managing the levels of light meant that his microscopes could achieve magnification as high as 275x, producing brighter and clearer images than any of his contemporaries.

Not only was van Leeuwenhoek a technical pioneer, he was also determined to explore the new worlds that these tools could reveal. He made lots of observations, becoming the first man in history to see many microscopic life forms, including bacteria, protists, and nematode worms. He discovered blood cells, the banded structure of muscle fibres, and he was also the first man to ever see his own spermatozoa swimming. He wrote numerous letters to the Royal Society in London, detailing his techniques, theories, and discoveries. He was initially met with some scepticism, as this microscopic world had simply never been seen before, but his findings were eventually verified by, amongst others, his hero Hooke. In 1680, van Leeuwenhoek was elected as a full member of the Royal Society.

Despite van Leeuwenhoek's incredible findings, microbiology was largely neglected after his death. There was no theory at the time that in any way connected these tiny micro-organisms with disease, as the prevailing wisdom was still the ancient idea that disease was caused by miasmas and bad air. As a result, the baton passed by van Leeuwenhoek was not taken up again for more than a century after his death, until the advent of Louis Pasteur and germ theory.

CHAPTER NINE

PUBLIC HEALTH

'By medicine life may be prolonged, yet death will seize the doctor too.'

William Shakespeare

Why did the Panama Canal take more than 30 years to build?

The Panama Canal is one of the largest and most challenging engineering projects ever undertaken. The canal links the Atlantic and Pacific Oceans, and it is one of the busiest shipping routes in the world. The project to build the canal was first begun in 1880 by France, but yellow fever and malaria decimated the workforce, killing huge numbers of labourers. Panama has a humid, tropical climate, making it an ideal breeding ground for mosquitoes, which transmit both malaria and yellow fever.

In response to the threat of disease, the French invested heavily in modern hospital facilities, but no measures were taken against mosquitoes, as no one yet knew that mosquitoes transmit disease. Most of the project buildings had no mosquito screens or nets, and hospital beds often had pans of water around their legs, which provided a perfect environment for mosquito larvae to grow. In 1893, the Panama Canal project was abandoned, after more than 52,000 workers had fallen sick, and perhaps as many as 20,000 had died.

In 1903, the project was restarted by the US, which bought out the French equipment and rights. The US approach was considerably better organised, under the leadership of engineer John Frank Stevens. The Panama Railway was rebuilt, to support the construction, and proper accommodation was built for the workers.

However, the most important change was a fresh approach to the problem of disease. In the intervening years, Sir Ronald Ross in India had conclusively demonstrated that malaria was transmitted by mosquitoes. Now, a war against the pests was fought by the army doctors Major Reed and the attractively named Colonel Gorgas. They oversaw a hugely ambitious and expensive eradication programme. Mosquitoes lay their eggs on the surface of standing water, so any pool or puddle that could be found was either drained, or sprayed with oil and insecticide. Small streams would have a dripping oil bucket placed over them, creating a film of oil over every still patch of water, which had the effect of killing the mosquito larvae. All canal workers were given free quinine, and all homes were fumigated, using up the United States' entire national stocks of sulphur and pyrethrum. Anyone who became infected was immediately quarantined, which meant they were kept in screened, mosquito-proof cages, thus preventing them from reinfecting. The campaign was enormously expensive – it was estimated to have cost as much $10 per mosquito killed.

Still, it worked. The canal was opened in 1914, two years ahead of schedule. In 1906, there was only case of yellow fever among the construction workers. From 1907 to 1914, there were none. Malaria was more persistent, but between 1906 and 1909, the death rate from malaria fell dramatically among employees, from 11.59 per 1,000, to 1.23 per 1,000. Over the period of

US control (1904–1914), there were only 5,609 fatalities, which marked a huge improvement. Today, the canal zone is free from both yellow fever and malaria.

Who was Typhoid Mary?

Mary Mallon was the first person in the United States to be diagnosed as a healthy carrier of typhoid. She was born in County Tyrone in 1869. After emigrating to New York in her teens, she became a cook, working for a series of wealthy families. Everywhere Mary went, people seemed to fall sick with typhoid, but she had never had the disease herself, so there seemed to be no reason to suspect that she herself could be responsible.

In 1906, she travelled to Long Island with the family of banker George Warren, to spend the summer cooking for them. After a few days of eating Mary's food, the family started to fall ill. An investigator, George Soper, was called, and he soon determined that the likely cause was Mary herself. Soper knew about the possibility of a person being a healthy carrier, meaning that they might carry a disease without ever having suffered from it themselves, and then transmit it to others who may suffer the full effects. In the case of typhoid, a healthy carrier may have suffered only mild flu-like symptoms when they first became infected, and then survive to pass on this fatal disease to others.

Typhoid fever has been a major killer for centuries, and it is still endemic in much of the world. It is caused by the bacillus Salmonella typhi, which can be transmitted in human urine and faeces. Symptoms include sudden and prolonged fever, nausea, diarrhoea, and headaches. Typhoid is usually passed on through contaminated food or water, particularly if the person preparing the food doesn't wash their hands after going to the bathroom. When questioned, Mary stated that she rarely washed her hands, as she felt there was no need.

Soper tried to approach Mary, to request samples of her urine and faeces for testing, but she reacted furiously, and lunged at him with a carving fork. After a few more attempts were met with similar aggression, the authorities were forced to intervene, and Mary was forcibly quarantined, kept in a cottage on Brother Island on the East River. This presented something of a legal and ethical conundrum, as Mary was being held prisoner without trial, against her will, having committed no crime. The concept of a healthy carrier was a relatively recent discovery, and there was no law that covered it. The Greater New York Charter did have a provision to prevent a person from wilfully infecting others, if that person posed a threat through being 'sick with any contagious, pestilential or infectious disease', but of course although Mary certainly did pose a threat, she wasn't technically sick.

While incarcerated, Mary continued to protest and appeal, as public interest in the case grew. Eventually she was released, on the condition that she sign an affidavit swearing to never again work as a cook. She agreed, but soon disregarded her promise once she was freed. Mary was single and childless, and none of the other jobs available to her paid as well as cooking did.

In 1915, typhoid broke out at the Sloane Maternity Hospital in Manhattan. 25 people were infected, and it soon became clear that the source was a cook named Mrs Brown, who of course turned out to be Mary working under a false name. When Mrs Brown's true identity was eventually revealed, public sympathy for Mary quickly drained away. It was clear by this point that Mary had been informed about healthy carriers, even if she didn't believe in them, and her use of a false name added to the impression that she was wilfully risking other people's lives. Eventually, Mary was quarantined again, and she remained incarcerated for the rest of her life, before eventually dying of pneumonia in 1938. The autopsy found that she was still infected with typhoid bacteria on the day of her death.

Over the course of her career, Typhoid Mary is known to have infected at least 53 people, killing three, but the numbers may well be much higher. Her name even lives on into the internet age, as a 'Typhoid Mary' is now a slang term for a computer user who recklessly refuses to install adequate virus and malware protection, and consequently endangers other computer users in their network.

What does Vitamin E do?

Bizarrely, no one really knows, and it may even do nothing, or at least nothing useful. Vitamin E is found in a wide range of foods, including nuts, seeds, wheat, eggs, spinach, avocado, and asparagus. For some time, the vitamin was thought to function as an effective antioxidant, which protected cells from ageing by combating free radicals. There were hopes that high doses of Vitamin E could help to prevent heart disease, cancer, strokes, and Alzheimer's disease.

However, recent clinical studies have been unable to demonstrate any of these benefits, and in fact have suggested that high doses of Vitamin E may be actively harmful. In 2005 the Heart Outcomes Prevention Evaluation trials, which studied more than 10,000 patients of 55 years old and over, found no benefits to the heart from taking high doses of Vitamin E, and in fact the vitamin seemed to be linked to a higher risk of heart failure. Other studies have also failed to find any benefit for Vitamin E in preventing heart disease.

The Heart Outcomes trial also looked at cancer, and found no difference between those taking Vitamin E and the control group, who didn't. Other recent studies have looked at cancer in women, and prostate cancer in men, and also found no benefit in taking Vitamin E. A National Cancer Institute study into lung cancer suggested that those taking Vitamin E were at a slightly higher risk of developing the disease. Other studies have looked into possible links with Alzheimer's Disease, but again found no benefit to taking Vitamin E supplements.

If anything, Vitamin E may be actively harmful, if taken in high doses. A meta-analysis conducted by Johns Hopkins University in 2004 found that the risk of dying within five years rose by about 5% in people taking at least 400 international units (IU) of Vitamin E per day. One possible explanation for this is that Vitamin E is an anti-coagulant, which may prevent blood clotting.

Why were low-lying towns particularly susceptible to cholera?

After John Snow's conclusive demonstration that the Soho out-break of cholera was caused by the contaminated Broad Street pump, one might imagine that the case for cholera being trans-mitted by water would have been resolved. Instead, however, there was considerable doubt about Snow's findings. Many scientists still believed that diseases were primarily caused by miasmas, and cholera was no exception. They felt that Snow's epidemiological map proved nothing, because the concentration of cholera cases around the pump only demonstrated that the ground by the pump, or somewhere near it, was the source of bad air.

In fact, Snow had already made a convincing case against such objections. His study had included a number of cholera victims who had taken water from the Broad Street pump, but who did not live in the area, and therefore would not have been affected by bad air. The most compelling example was a woman who lived in Hampstead, an area which was untouched by chol-era, who never even went to Broad Street herself, but who was regularly supplied with a bottle of water from area, because she had lived there in the past, and liked the taste of the water. She developed cholera, as did her niece who drank from the same bottle.

Snow's chief opponent was William Farr, superintendent of the statistical department at the General Register Office, who was widely recognised as the leading authority on the use of statistics in the study of disease. Farr had produced a persuasive study, comparing the rate of cholera with elevation from sea level. Amazingly, Farr's study seemed to clearly show that the closer a person lived to sea level, the more likely they were to contract cholera. This, Farr argued, was clear evidence that the source of infection was something in the ground.

However, Snow responded by pointing out that the most elevated towns in Britain – Wolverhampton, Dowlais, Merthyr Tydvil, and Newcastle-upon-Tyne – had all suffered enormously from cholera. Whereas, many low-lying sites which happened to be supplied by a well, such as the Queen's Prison, Bethlehem Hospital, and Horsemonger Lane Gaol, had all largely remained untouched, despite being surrounded by areas which were badly affected.

Snow demonstrated that the real issue was not elevation from sea level, but where towns drew their water from. The only reason for the correlation between elevation and cholera was because low-lying towns tended to draw their water from tidal rivers, which meant that their water supply was liable to become infected by sewage travelling upstream.

By 1866, Farr had become a convert to the cause of germ theory. He produced a paper demonstrating the increased mortality rate for people who drew their water from the Old Ford Reservoir in East London. Thanks to the work of Edward Jenner and Louis Pasteur, germ theory was becoming widely accepted, and major cities were building new facilities to collect and treat sewage, a process which ultimately resulted in cholera being eradicated in Western cities.

Was the 'father of jogging' killed by jogging?

Jim Fixx was the key figure in inspiring the jogging boom which began in the United States in the late 1970s, and arguably continues to this day. He wrote several best-selling books, including 'The Complete Book of Running', which had a major impact on its release in 1977, selling more than a million copies. Fixx, who has been described as the 'founding father of jogging', had started running in 1967, when he weighed 240 pounds, and was a heavy smoker. He credited jogging with transforming his health, helping him to lose 60 pounds and give up smoking, extending his life.

On the July 20, 1984, Fixx died of a heart attack, at the age of just 52. He was jogging at the time, which has led to the widespread belief that his death was caused by jogging, and that therefore the health benefits claimed for jogging are illusory. Famously, the scathing comedian Bill Hicks, who had little time for health fanatics, developed a routine mocking Fixx, and relishing the irony of his death.

However, the evidence suggests that Fixx's death was not in fact caused by jogging. His family had a history of heart trouble; his father had suffered a heart attack at 35, and had died of a second one at the age of just 43. Jim Fixx's autopsy showed that he died of blocked arteries: three of his coronary arteries were blocked with cholesterol, suggesting that a more likely risk factor for his death was diet, not jogging, perhaps linked to his earlier, unhealthy years before he took up jogging.

It it true that the risk of a heart attack goes up slightly during exercise, but this risk is much lower overall for people who exercise regularly, and pales into insignificance compared with the enormous benefits that exercise brings. In just one example, numerous studies have shown that regular exercise reduces the risk of a heart attack by between 50% and 80%. The likelihood therefore is that, while he may have died relatively young, Jim Fixx's life was nonetheless extended by jogging.

Why are pig farmers particularly likely to have their appendix removed?

In 1991, a Finnish scientist called Markku Seiru conducted a fascinating study into pig farmers and abattoir workers in Finland, comparing the rate of appendectomies within this group with that of the general population. The study found that pig farmers were around two and a half times more likely to have their appendix removed than other Finns, while abattoir workers were almost four times more likely to have an appendectomy. But why, you may wonder, should working with pigs carry such a bizarre and specific risk?

The answer is that a high proportion of pigs carry a bacterium called Yersinia enterocolitica, which can contaminate food, and causes a disease called yersiniosis. Yersiniosis is usually caused by eating undercooked meat, unpasteurised milk, or contaminated water. The symptoms include diarrhoea, fever, and abdominal pain, and it is often confused with appendicitis. If properly diagnosed, doctors will often allow an infection to run its course, while even severe cases are usually treated with nothing more drastic than antibiotics.

However, because these symptoms often resemble appendicitis, it is not uncommon for patients to be misdiagnosed, and given unnecessary appendectomies. At the time of the Finnish study, around 35% of pigs in Finland were found to carry Yersinia enterocolitica, which presumably explains the high rate of appendectomies among farm workers. Because this group were much more likely to encounter yersinia bacterium, they were therefore also more likely to have their symptoms confused with appendicitis, leading to unnecessary appendectomies.

Why did organ donor numbers drop suddenly in 1978?

In 1978, there was a sudden, dramatic drop in the number of potential organ donors coming forward, with the reduction reportedly as being as high as 60%. Officials were puzzled at first, as there seemed to be no obvious reason why rates should drop so suddenly. Then, a compelling theory emerged, which suggested that the cause of this decline was a bestselling novel called 'Coma', written by Robin Cook, which in 1978 had been adapted into a hit film directed by Michael Crichton. 'Coma' was a creepy medical thriller which told the story of an unscrupulous hospital where patients were deliberately killed, so that their organs could be harvested and sold to rich clients. Both the book and the film made a big impact, and after their release, rates of organ donation dropped dramatically.

Thankfully, the public's sensitivity to stories about organ donation can also be a force for good. In 1982, the parents of 11-month old baby Jamie Fiske were desperate. Their daughter had been born with biliary atresia, a congenital liver condition which at the time usually resulted in death before the age of four. The family's only hope was a liver transplant, but there were no suitable livers available. Luckily, Jamie's father was a well connected hospital administrator, and so he instigated a massive publicity campaign, contacting thousands of doctors, and recruiting major public figures such as senator Edward Kennedy and news anchor Dan Rather, to lobby on Jamie's behalf. The campaign worked. Jamie Fiske became a major news story, and the families of 500 potential donors contacted the family, offering to help. A suitable liver was found, and Jamie survived.

Even more importantly, thanks to all this publicity, requests for organ donor cards went up by more than 30%, and the number of families allowing doctors to remove organs from relatives who suffered irreversible brain damage went up from 50% to 86%.

Are we living longer?

The simple answer is of course 'yes', although it is slightly more complicated than it might first sound. Average life expectancy has been increasing for centuries, with a dramatic rise since the 19th century. In medieval Britain, average life expectancy was around 30 years. In the centuries that followed, any rise in life expectancy tended to be gradual, and this figure is not thought to have climbed above 40 at any point until the start of the 20th century, when it increased dramatically. Over the course of the 20th century, life expectancy in many parts of the world doubled. In the US the current life expectancy at birth is 78, and in many parts of the world it is over 80. These increases are chiefly attributed to improved sanitation and clean water, as well as improvements in medicine, nutrition, childbirth, surgery, and the development of vaccines.

However, the definition of life expectancy may warrant some explanation. In ancient Rome, the average life expectancy at birth was 25, but this is not to suggest that most people died at around the age of 25. 25 was an average, taking into account the huge number of people who died as babies, in a period when infant mortality was extremely high; or who died young from childhood diseases, or military conflict, and so on. Anyone who managed to survive their early years would see their life expectancy increase dramatically, even as they got older. Thus, while the average life expectancy in Roman times may have been 25, many people did live on to what we would consider to be a ripe old age.

In fact, throughout history, many people have lived to 90 and beyond, and perhaps even into the hundreds. In ancient Egypt, the Sixth Dynasty pharaoh Pepi II is believed to have lived beyond 100. Pharaoh Ramesses II was recorded as having lived to about 90, and tests that were carried out on his mummified tomb seemed to confirm this. In ancient Greece, the philosophers Pyrrho, Eratosthenes, and Xenophanes all reached their nineties. Socrates died just before his 90th birthday, and only then because he was executed. The astronomer Hipparchus of Nicea is said to have lived to 109. In the fourth century AD a number of Christian priests lived into their 90s. Michaelangelo was still working at 89, and a number of Native American chiefs lived into their hundreds.

In other words, at many points throughout human history there were lots of people living longer than the vast majority of people living today. The oldest person to have lived was Jeanne Calment of France (1875–1997), who died at the age of 122. This suggests that, despite all our advances in medicine, sanitation, nutrition, and health, the maximum human life span doesn't seem to have increased by a huge amount, even though the average life expectancy certainly has. This is demonstrated even today, in parts of the world where medicine has had only a very limited impact. The lowest life expectancy rates in the world are found in Africa, and yet Africa also produces centenarians. In regions of the Caucasus, there are said to be much higher rates of centenarians than in the US, and modern medicine has presumably had only a limited impact there too.

Another interesting point is that, while average life expectancy has clearly increased dramatically, it's not clear how much longer this trend will continue, and some suspect that it may already have peaked. Demographers agree that there will be many more centenarians in the decades to come, and yet life expectancy at birth may actually be beginning to drop, at least in the US. Today, a much higher proportion of the population are overweight or obese, particularly among the young, and many people tend to lead sedentary lives, with little exercise. The majority of deaths in advanced Western countries are caused by heart disease, cancer, diabetes, and stroke, and those with sedentary lifestyles and high BMIs are more at risk from these diseases. As a result, a recent report suggested that average life expectancy could drop by as much as 5 years, based on current rates of obesity.

Who was Patient X?

Patient X was the name given to a former US Marine who preferred to keep his true identity a secret, after inadvertently taking part in what was to become an award-winning medical study. He was bitten on the lip by his pet rattlesnake, and insisted on curing the venomous injury himself, using electro-shock therapy. He did this by attaching the spark plug wires from his car to his lip, and revving the engine to 3,000 RPM for five minutes.

Thanks to this courageous, albeit somewhat foolhardy experiment, Patient X was the joint winner of the 1994 Ig Nobel prize for medicine. He shared the prize with Dr. Richard C. Dart of the Rocky Mountain Poison Center and Dr. Richard A. Gustafson of The University of Arizona Health Sciences Center, for their 'well-grounded' medical report 'Failure of Electric Shock Treatment for Rattlesnake Envenomation'. After receiving his award, Dart commented, 'I was stunned to receive the 1994 Ig Nobel Prize in Medicine, although not as shocked as our patient.'

The Ig Nobel awards are a series of light-hearted science awards, whose stated aim is, 'To first make people laugh, and then make them think.' In 2009, the winner of the Medicine prize was Donald L. Unger of California, who had cracked the knuckles of his left hand twice a day for 50 years, while rarely cracking the knuckles of his right hand, to test whether or not knuckle-cracking causes arthritis. He found that after 50 years of this experiment, there was no arthritis in either hand, and no difference between the two hands, and thus concluded that knuckle-cracking does not cause arthritis.

Did the Yankee Stadium have to fit wider seats, to accommodate the increasing girth of fans?

Indeed it did. The original Yankee Stadium was built in the 1920s, when the average American was considerably thinner. At that time, seats were just 15 inches wide, but as the American population have grown steadily wider, this proved to be uncomfortable for all concerned. Since 1980, obesity rates in the US have doubled, and today around 65% of Americans are overweight or obese. In response to this trend, the owners of the Yankee Stadium removed 9,000 seats, and replaced them with seats which were considerably larger, at 19 inches wide. In 2009, a brand new Yankee Stadium was opened, just across the road from the old site. Again, seat sizes were increased. In the new stadium, seat widths range from 19–24 inches.

Nor is this a problem restricted to the United States. In the UK, more than half the population are now overweight or obese, and the rate of obesity has increased almost fourfold in the last 25 years. In London, plans for the 2012 Olympic Stadium had to be redrafted to accommodate the ever-increasing size of British bums. In the original plans, all 20,000 seats were to be 18 inches wide, but they will now be increased by 2 inches. British amusement parks have also started offering special rows of 'outsized' seats for larger visitors.

This issue has been a particular problem for airlines. Airlines will generally try to cram in as many seats as possible, to maximise their earnings, and keep prices competitive. There was a recent outcry when British Airways tried to cram another line of seats into the economy section of its Caribbean flights, making each row ten seats wide, rather than the usual nine. In the end, the company received so many complaints that it reversed the plan. In 1996, the Civil Aviation Authority published a report recommending a minimum seat width of 19.6 inches, but some airline seats are just 16.2 inches wide.

As a result, many obese passengers simply can't fit into a single seat. Many airlines now encourage or require larger passengers to buy a second ticket, so that they can sit on two seats, a policy which has caused some controversy. In February 2010, filmmaker Kevin Smith reacted furiously when he was removed from a Southwest Airlines plane prior to departure, apparently because he was considered too fat to fly safely in just one seat.

Overweight passengers feel that this kind of policy is discriminatory. Others argue that it's only fair that each passenger should pay for the space they take up, citing horror stories of thin passengers being crushed by portly co-passengers. In 2002, Virgin Atlantic paid £13,000 compensation to Barbara Hewson from South Wales, who suffered sciatica, torn leg muscles and a blood clot after being crushed by an obese co-passenger on a transatlantic flight.

Why are there so many contradictory health stories in the news?

Every week, the news media seem to bring us a new health breakthrough or scare, and as often as not it completely contradicts a different health story from only a few weeks ago. These stories often focus on food. Chocolate, we're told, is an excellent source of beneficial antioxidants, but then next week it is a key factor in childhood obesity. Red wine lowers the risk of heart attack, but then the next week it causes liver disease. Coffee lowers the risk of heart disease, but then caffeine can increase blood pressure, increasing the risk of a heart attack. Oily fish can boost brain function and IQ, but they also contain poisonous mercury, increasing the risk of heart disease. Soy milk is healthier than cows' milk, except that it increases the risk of breast cancer. And so on.

The truth is that no one foodstuff can magically resolve all our potential health issues. The best way to look after your body is simply to follow the basic rules for a healthy lifestyle that we all generally know: avoid smoking, take regular exercise, eat a balanced diet including fruit and vegetables, drink in moderation if at all, and get plenty of sleep.

So why do these contradictory health stories keep appearing in the press? The main reason is that they make good stories, which people love to read. They tend to contradict one another simply because the studies these stories are based on rarely draw the kind of dramatic conclusions that the media claim for them. Studies of this type are often run on a small scale, and are merely intended to test a possible mechanism, before more detailed research takes place. A hypothetical case-control study might look at coffee drinkers, for example, and find that they have less heart attacks than the rest of the population. This doesn't prove that coffee reduces the risk of heart attacks, it merely suggests that there might be a possible link, and that further studies should now take place. Hopefully, there will then be more studies, using control groups, and randomisation to eliminate sources of bias, in an effort to make all things equal except for the subjects' coffee drinking habits.

Even then, the study is pretty much bound to be unsatisfactory in some way. For a study to be really effective, it needs to sample a large number of people, with controls and randomisation, and follow them for a very long time – after all, the effects of coffee on the heart may take years to have any effect. Big studies of this kind are extremely expensive and difficult, as well as carrying ethical implications. If the researchers really believe that there is a likely link between coffee and heart attacks, is it justifiable to ask the control group to not drink coffee for years, possibly endangering their health in the process?

Big studies of this kind are rare, so health stories will often imply a causal link where none has been shown. For example, a number of recent studies have shown a possible correlation between eating processed pork and higher rates of colorectal cancer. The press have reported this hysterically with headlines such as the Daily Mail's 'Why eating just one sausage a day raises your cancer risk by 20 percent'. However, correlation does not imply causation. There are many reasons why the kinds of people who eat processed pork might also suffer from higher rates of bowel cancer.

More worryingly, these stories are sometimes based on studies which never even appear. Newspapers will be sent an exciting-looking press release, and run a sensational story on the back of it, but then the actual study never materialises in any scientific journal. For example, in February 2004 the Daily Mail ran a story about a forthcoming study that demonstrated that cod liver oil was 'nature's superdrug', based on a press release from Cardiff University. However, a year later, no study had been published.

Who is 'Frozen Dead Guy'?

In 1989, a Norwegian man named Trygve Bauge brought the frozen corpse of his grandfather Bredo Morstøl to the United States, where he intended to keep it cryogenically frozen. Trygve was a passionate believer in cryogenics, also known as cryonics, which is the practice of freezing dead bodies to preserve them, in the hope that science will one day be able to bring them back to life. Trygve was not only trying to save his grandfather for the future, he planned to set up his own cryonics clinic, in Nederland, Colorado where he set up home.

However, Trygve's plans were in truth somewhat amateurish. If you imagine a cryonics institute, you're presumably picturing some kind of shiny, high-tech chrome and glass laboratory facility, with gleaming, carefully engineered caskets submerged in liquid nitrogen. Trygve's facility, on the other hand, consisted of a single, run-down shed, which contained a coffin in a plywood box, filled with dry ice.

Sadly, Trygve had to give up on his ambitious plans when he was deported in 1993 for overstaying his visa. He left behind his mother, Aud, and his grandfather's frozen body. When the local authorities found out about Bredo's frozen corpse, which was now at risk of thawing, their first reaction was to pass a law banning the keeping of frozen bodies. This law, incidentally, also made it technically illegal for anyone to keep a frozen chicken in their freezer, but no one seemed to be all that bothered about enforcing that particular interpretation of the law. Nonetheless, the passing of the law raised more local interest, and Bredo was starting to generated significant press coverage. The local authorities decided that an exception could be made for Bredo's frozen corpse, even though this corpse was the sole reason the law had been passed in the first place, and so Bredo was allowed to stay. Trygve arranged for a local company to continue regularly replacing the dry ice, and another local company even built the family a new shed.

Over time, the town became famous for its unusual resident, and so decided to capitalise on this, by setting up an annual celebration. The first full weekend of March became the 'Frozen Dead Guy' festival, bringing 5,000 tourists into this remote town for a series of bizarre festivities. There are coffin races, along a course which goes directly over the roof of Grandpa Bredo's shed. There are snowshoe races, and snow sculpture contests, and a Grandpa Bredo lookalike contest. There is Frozen Dead Guy ice cream, and champagne tours of the famous shed. There is also a dance, with the slightly juvenile name 'Grandpa's Blue Ball', at which revellers come dressed as corpses. Hardy visitors can even attempt the 'polar plunge', a dive into an ice-cold reservoir, but you usually have to break through the ice first.

However, one of the downsides of setting up your cyronics facility in a shed is that the technical challenges are significant. Grandpa Bredo's corpse does not seem to have been kept to the very highest cryogenic standards, and has almost certainly defrosted more than once. This sadly means that even if it were to become possible for scientists to revive a frozen corpse, there seems to be little chance of a successful resurrection for Colorado's very own Grandpopsicle.

What was a 'London Particular'?

A London Particular was a kind of thick smog that affected London intermittently for around 100 years, from the Industrial Revolution until at least the 1950s. Smog is a portmanteau word, produced by combining the words 'smoke' and 'fog', as smog was caused primarily by exhaust fumes and industrial byproducts, in particular the burning of coal.

During this period, London Particulars were a common and routine occurrence. They were known colloquially as pea-soupers, in reference to the yellow tinge that the fog often took on, thanks to its high sulphur content. Amusingly, in a strange backwards twist, pea soup itself became known as 'London Particular', in reference to the famous smog.

Although pea-soupers were common, they were nonetheless enormously disruptive. An incredulous Time magazine article from 1951 describes the following absurd scene, which occurred on a runway at Heathrow Airport (which at the time was simply called London Airport). First, the plane lands in a thick fog, and is ordered to stay in position, as visibility is too poor for the pilot to taxi back to the arrival terminal. A bus is sent out to pick up the passengers, but it gets lost in the smog. A truck is then sent out to find the bus, but it too gets lost. Soon, there are five separate search parties out searching, hopelessly groping around in the fog. Eventually, a man on a motorbike finds the plane, but realises he now has no idea where the terminal is, so he rides off. Eventually, the passengers are rescued, but when they get to the terminal, their luggage has been lost.

The smog that appeared in December 1952 was unusually thick, but at the time it didn't seem to be anything particularly out of the ordinary. However, a number of unusual factors were about to combine to spectacular effect. The weather had been very cold, which meant people had been burning more coal than usual to keep warm. There was also more pollution from exhaust fumes, as all the trams had recently been replaced with diesel-spewing buses. Additionally, prevailing winds from Europe had brought industrial pollution across the channel, which now became trapped in an anticyclone, with a layer of cold air trapped under a blanket of warm air.

As a result, this smog was thicker and longer-lasting than it had been on any previous occasion. Visibility was limited to just a few yards, making driving practically impossible. Buses and ambulances stopped running, and the only working public transport was the London Underground. As a result, there was no way for people to get to hospital. Amazingly, the smog even managed to find its way indoors, and concerts and film screenings were cancelled because the audience couldn't see what was happening on the stage and screen.

After the smog subsided, things seemed to quickly return to normal. However, soon there were a number of strange indications. Undertakers found they were running out of coffins, and florists ran out of flowers and wreaths. There were no bodies littering the streets, but without anyone realising it, the death rate had soared, to three or four times the normal level. Most of the fatalities were caused by bronchitis and lung infections. 'Smog' was not traditionally listed as a cause of death, so statistics are hard to come by, but estimates suggest that as many as 12,000 people died as a result of just five days of smog, with around 100,000 more falling ill.

If any good can be said to have come from such a disaster, it is that the public outcry that followed the Great Smog led to the implementation of new laws to protect the environment and cut air pollution. The City Of London Act of 1954 and the Clean Air Acts of 1956 and 1968 restricted the types of fuel that could be used in industry, leading to a huge improvement in air quality. From our perspective today, as we face the potential disaster that is climate change, examples such as this set a powerful precedent. With the right combination of regulations and incentives, London Particulars, Los Angeles Smog, and 'acid rain' have all become outdated, forgotten terms, raising hopes that one day we may feel the same combination of curiosity and nostalgia for phrases like 'carbon footprint' and 'greenhouse gas'.

What is the Cape Doctor?

The 'cape doctor' is the name given by locals to the strong prevailing wind found on the Cape of Good Hope on the South African coast between spring and late summer. It is known as the Cape Doctor because for centuries the wind has been believed to carry pollution and pestilence away from Cape Town and out to sea. The Cape Doctor can be extremely strong – it can blow double decker buses over, and have people clinging horizontally onto lampposts for dear life.

It might strike you as odd to give a name to a wind system, but in fact there are lots of other prevailing winds from around the world which have their own specific name. You may for example have heard of the Mistral, which is a strong, cold wind that blows from northern France down to the Mediterranean. There is also another wind doctor, the Fremantle Doctor, which is the name for the cooling sea breeze in Western Australia, which may have gained its name by blowing away the stench of burning corpses which blighted the ailing colony in its early years.

However, my personal favourite is the Williwaw, which is a sudden blast of wind descending from the coastal mountains at the Strait of Magellan, the waterway which divides the southern tip of South America from the archipelago of Tierra del Fuego. Historically, this was the main shipping route between the Pacific and Atlantic oceans, before the Panama canal was built. However, thanks to the unpredictable and powerful Williwaw, it was also one of the world's most dangerous stretches of water.

CHAPTER TEN

SNAKE OIL

'God heals, and the physician has the thanks.'
Traditional proverb

Why is a quack called a quack?

A quack is someone who fraudulently claims to have medical expertise of some kind, and the term is often used to refer to the sellers of patent medicines. The name 'quack' has nothing to do with ducks; it comes from the archaic Dutch word 'quacksalver', which means 'a boaster who applies a salve'. A quacksalver would thus quack about his salves, which is a pretty good definition of a travelling medicine man, selling his dubious nostrums at markets and county fairs.

Does 'snake oil' work?

When the words 'snake oil' are used today, they are generally meant as a euphemism, a synonym for fraudulent quackery. This use of the term refers to a bygone age, when patent medicine men travelled from town to town, selling snake oil, vegetable tonic, and Indian root pills. These remedies rarely worked, and rarely even contained the ingredients they claimed to. The skill of these travelling 'doctors' was all in the marketing, filled with extraordinary, bogus pseudo-scientific claims and anecdotes. These hucksters would also frequently use shills, accomplices in the crowd who would 'volunteer' for demonstrations, then pretend to be instantly, dramatically cured by the remedy in question. Today, we use the term 'snake oil' to refer to pseudo-science, absurd health claims, and any kind of exaggerated salesmanship.

However, even though the name 'snake oil' has become synonymous with fraudulent quackery, it seems that the oil itself might, astonishingly, work. Snake oil has been used for centuries in traditional Chinese medicine, where it is known by the name *shéyóu*. It has traditionally been used as a pain reliever and anti-inflammatory, to relieve joint pain, rheumatoid arthritis, and bursitis, and in fact it is still sold and used today. The snake oil in question comes from a particular type of snake called the Chinese Water Snake, a species found in China, Taiwan, and Vietnam.

Snake oil (by which I mean genuine snake oil) was probably brought to America by the gangs of Chinese labourers who worked on the Transcontinental Railroad, the railway which was built in the 1860s to link the east and west coasts. This was backbreaking work, and the Chinese workers are said to have rubbed snake oil into their aching joints to ease the pain.

We know today that Chinese water snakes are a rich source of eicosapentaenoic acid, also known as EPA, which is an omega-3 fatty acid. Omega-3 fatty acids are found in fish oil, and are believed to bring about a range of health benefits. EPA in particular is known to be an effective anti-inflammatory and pain reliever. In other words, despite its dubious reputation, snake oil might actually work.

However, this raises a rather tricky objection. If snake oil was one of the few patent medicines, or perhaps even the only one, which actually worked, how did it acquire such a bad name? The answer may be found in the specific composition of the products being sold. As mentioned above, Chinese water snakes have very high concentrations of EPA in their bodies, around 20% of their oil is EPA. The oil of American rattlesnakes, on the other hand, contains only 8.5% EPA. Even if the travelling hucksters were selling a product containing genuine snake oil, it would rarely if ever have come from a genuine Chinese water snake, so the odds are that it would contain little or no EPA. One of the most popular American snake oil products was Stanley's Snake Oil, produced by Clark Stanley, the self-proclaimed 'Rattlesnake King'. Tests in 1917 found that Stanley's Snake Oil was made up of turpentine, camphor, and mineral oil – but no snake oil. In other words, genuine snake oil may have actually worked, but the product that was sold would very rarely have been genuine.

What was the Doctor's Riot?

Right now, you might be imagining an angry mob of men in white coats, wielding clipboards and stethoscopes, but in fact this was not a riot *by* doctors, but rather a riot *against* doctors. According to one report, the trouble started at the New York Hospital in April 1788, when a medical student named John Hicks Jr. waved the arm of a corpse at a group of nosy children who were peering into the dissection room. 'This is your mother's hand,' he shouted. 'I just dug it up. Watch it or I'll smack you with it!'

The children ran off, but one boy was particularly upset, as his mother had recently died. Could it really have been her arm? He went home and told his father what had happened, and they then went to inspect her grave. Sure enough, the grave was empty, having recently been dug up, and whoever had done it had smashed open the casket, and hadn't even bothered to refill the grave with earth. The husband was furious, and vowed that someone would pay for this outrage. He gathered together a mob of his friends, and they marched through the streets of Lower

Manhattan to the New York Hospital, with the crowd swelling to hundreds along the way.

There was a reason for this outpouring of public outrage. Grave-robbing was a growing problem, as medical students and anatomy teachers needed fresh corpses for their studies. They usually only took corpses from the graves of the poor, homeless, or black, but in times of great demand, any grave would do. The problem had become so bad that wealthy families would post armed guards at the graves of their relatives for two weeks after burial. After two weeks, the body would be too decayed to be any use for dissection.

The angry mob surrounded the hospital, blocking all the exits, and baying for the doctors' blood. Luckily, most of the doctors had already managed to escape through the hospital's rear windows. One bravely stayed behind, along with three medical students, to guard the precious instruments and anatomical specimens, but it was to no avail, as the rioters broke in and destroyed everything in sight.

The mob were still not satisfied, so they then headed out into the streets, looking for doctors to attack. One innocent man was beaten to the ground, simply for being dressed in black. The home of Sir John Temple was completely destroyed, despite him having no connection to medicine – it seems that the rioters had misread 'Sir John' as 'surgeon'.

The riot continued the next day, with the mob attacking Columbia College, destroying valuable medical specimens and tools. That evening, the epicentre moved to the Manhattan Jail, where a number of doctors were being held for their own safety. The militia guarding the jail was led by Baron Friedrich von Stueben, who refused to use force, until one of the rioters hit him with a hurled brick, at which point he changed his tune, crying 'Fire, Governor, fire!'. At least five of the rioters were killed, with many more seriously wounded. The doctors treated the injured, and the riot ended.

Weeks later, a law was passed in New York allowing for the dissection of hanged criminals, but this made no difference, as the demand was so high that grave-robbing continued unabated, until at least the 1850s. And although a riot against doctors might sound like a strange and isolated incident, in fact there were as many as 25 similar incidents of public outrage at medical schools throughout the US up until 1884.

How did Dr Alpheus Myers' 'tapeworm trap' work?

In 1854, Alpheus Myers, a rural doctor from Logansport, Indiana, filed a patent for an ingenious new product, a 'new and useful Trap for Removing Tapeworms from the Stomach and Intestines.' The trap consisted of a spring-loaded cylinder of gold, platinum, or some other rustproof metal, about three quarters of an inch long, and a quarter of an inch in diameter, attached to a cord. The cylinder would contain the bait; according to the doctor, cheese would be an attractive bait for a hungry tapeworm.

Before using the trap, the patient would have to fast for a week. The idea of this was to starve the tapeworm, to force it to climb up into the stomach, or even the throat, in search of food. After fasting, the patient was instructed to swallow the trap, leaving one end of the cord hanging from his mouth, and then wait for six to twelve hours. During this time, Dr Myers predicted that the tapeworm would take the bait, causing the spring-mounted cylinder to close around it. Myers warned that the spring should be strong enough to grasp the worm, but not strong enough to cut its head off. Myers believed the patient would notice when the tapeworm became trapped, either because they would feel it, or because the worm would tug on the cord. Then, 'The patient should rest for a few hours after the capture, and then by gentle pulling at the cord the trap and worm will with ease and perfect safety be withdrawn.'

As daft as it may sound, it's reported that the device may have been successfully used at least once. A year after the patent was granted, the Scientific American reported that Dr Myers had used the trap to remove a 50-foot tapeworm from a patient, 'who, since then, has had a new lease of life.'

What was Graham's Celestial Bed?

James Graham was a self-styled sex guru, and one of the leading quacks of the 18th century. He was born in Edinburgh in 1745, where he trained in medicine, although he never actually finished his degree. While travelling around America, where he presented himself as an eye specialist, he learned about electricity – from Ben Franklin, he claimed, but this was almost certainly false.

In 1775, he moved to London, where he set up an elaborate 'Temple Of Health' at the newly built Adelphi. At the Temple, health-seeking visitors could enjoy wandering through the ornate, perfumed rooms, listen to the orchestra, or hear Graham deliver lectures about health, nutrition, and electricity. Sometimes Graham would end his lectures by giving the audience a surprise jolt of electricity, through a wire attached to each seat.

Graham does not seem to have been much of a doctor, but he was quite a showman. His lectures were adorned with scantily-clad young women, his 'goddesses of health', one of whom was reportedly Emma Lyon, who went on to become the notorious Lady Hamilton. He won many fans among the great and the good including the politician Charles James Fox, and the Duchess of Devonshire, who turned to Graham for help after failing to conceive a child.

The centrepiece of Graham's temple was the Celestial Bed, which cost the enormous sum of £50 a night, and according to Graham was guaranteed to cure sterility or impotence. The bed was twelve feet long by nine feet wide, with powerful magnets underneath, and it could be tilted and adjusted to reach the ideal angle for conception. The mattress was said to be stuffed with rose leaves, flowers, and hair from the tails of fine English stallions. A large mirror looked down from the ceiling, and as the orchestra played nearby, perfume (and possibly nitrous oxide) wafted down, and electricity crackled across the headboard, on which were inscribed the words, 'Be fruitful. Multiply and Replenish the Earth.'

The Temple was a great success, and soon moved to a new site at Pall Mall. Graham became a well-known public figure, although many dissenters felt that he was a charlatan, whose use of sex and titillation were unseemly. He was also not much of a businessman, as despite its popularity his Temple consistently lost money, and he ended up deep in debt. He eventually returned to Edinburgh, where he began advocating the reviving properties of mud baths. Presumably, he no longer had the capital to invest in more expensive panaceas.

How did John Brinkley make his fortune?

John Brinkley was a remarkable quack, who became one of the richest and most celebrated doctors in America, despite having no medical qualifications. He was born in 1885 into relative poverty in North Carolina, and his childhood ambition was to become a doctor. At the age of 22, he began medical training, albeit at a dubious, unaccredited medical college, but he never completed it, as by this time he had mounting debts, and a young family to support.

In the course of his studies, however, he did learn about hormones, glands and their effects on the body, which would prove to be an important discovery. By 1918, Brinkley was 33, and he had spent more than ten years as a travelling doctor and patent medicine man. He now set up a large-scale clinic in the small town of Milford, Kansas, offering a very unusual treatment, which he marketed aggressively as a miracle cure for impotence, infertility, and other sexual problems. Brinkley claimed he could cure all such ills by surgically implanting goats' testicles into his patients. The theory was that the vigour and vitality of the randy goat would naturally revive the sexual appetite and capacity of the patient.

The procedure was a fairly simple one. The patient would check in to the 'hospital', having paid a considerable fee of $500 or more, the equivalent of around $5,000 today. The patient would then be led out to the back of the building, where he or she would select a lusty looking goat, which would then be castrated. Male patients would then have an incision made in their scrotum, into which the goats' testicles would be inserted, after which the cut would be sutured. Female patients had the goat testes implanted into their abdomen, somewhere vaguely near the ovaries.

This extraordinary treatment made Brinkley a huge fortune over the course of his life, thanks chiefly to his skills as a marketer and self-publicist. He was in almost constant conflict with the American Medical Association, who denounced him as a charlatan, but he bought his own radio station, which he filled with advertisements for his own products, clinics, and pharmacies. When this station was shut down by the US authorities, he moved across the border into Mexico, setting up the world's most powerful radio transmitter, allowing him to broadcast into the US with impunity. It was said that Brinkley's broadcasts could be heard as far away as Finland and the Soviet Union, and the radio signal was so strong that it illuminated car headlights, turned wire fences into receivers, and bled into private phone calls.

Obviously, there was no actual scientific merit to Brinkley's goat gland surgery, on the contrary it was extremely dangerous. Many of Brinkley's patients did believe they had been cured, but this was surely only a result of the placebo effect. Many others died, either because of a reaction to the rotting goat testicles that had been implanted inside their bodies, or because of gangrene or infection caused by Brinkley's unsanitary, unskilled surgical procedures. A recent biography (the excellent 'Charlatan' by Pope Brock) suggests that Brinkley, who was well aware that he was a fraud, could be regarded as one of America's most prolific serial killers, with many hundreds of victims.

At the height of his fame, Brinkley was a man of great standing. He ran for political office a number of times, and was only prevented from winning by the dubious electoral practices employed by his opponents. He was a multi-millionaire, with a fleet of cars, yachts, and an airplane, who even loaned one of his yachts to the Duke and Duchess of Windsor. His enormous influence also had some strange indirect effects. Audiences were unlikely to listen to a radio station which only consisted of adverts, so Brinkley filled the airtime with local Southern singers and performers, and as a result kickstarted the boom in blue-grass and Country and Western music, establishing Nashville as its home.

Brinkley's life came to a sad conclusion. He was pursued for decades by Morris Fishbein, the editor of the American Medical Association journal. The dispute eventually ended in court, in a case which Brinkley lost, opening him up to hundreds of lawsuits for fraud, malpractice, and wrongful death. By 1941 he was bankrupt. A year later, he was dead, following a series of heart attacks. One obituary described him somewhat euphemistically as 'the most fabulously successful medical maverick the country ever saw'.

What is a toad-eater?

Toad-eaters had one of the most unpleasant jobs in the history of medicine. A toad-eater was an assistant to a quack medicine man, whose job would be to publicly swallow a toad, as part of the medical demonstration. At the time, toads were believed to be deadly, and many species of toad extrude a milky, poisonous substance when threatened. The toad-eater's act was therefore considered to be tantamount to suicide. Having swallowed the poor amphibian, the toad-eater would collapse dramatically, contorting his body in agony. However, the mountebank (so called because he would climb onto a bench to sell his wares) would then pour some of his rejuvenating miracle nostrum into his assistant's mouth, at which point the toad-eater would be miraculously cured. On a good day, this spectacle would be enough to convince several in the audience to part with their cash, and the charlatans could move on to the next town.

Even at the time, it wasn't clear whether or not toad-eaters did actually eat the toads. There are lots of records of people having heard of someone who had once seen it happen, but first-hand accounts are hard to come by. Presumably, the toad-eater would often palm the toad, and simply pretend to eat it. Alternatively, they may have used a non-poisonous frog. Alternatively, it's possible that they did swallow the toad, and simply accepted the resulting few days of illness as the cost of doing business.

The practice of toad-eating is now largely forgotten, but it has had one enduring influence, as it gave us the word 'toady'. A toad-eater was someone who would risk illness and even death for his boss, which suggests he would have been an unusually lowly, supine, obsequious type of person. Thus, a toady today is a sycophantic employee who will suck up to their boss, and suffer any humiliation on their behalf.

Who was the Münchhausen of Münchhausen syndrome?

The answer is Karl Friedrich Hieronymus, Freiherr von Münchhausen, a German baron who was born in Bodenwerder in 1720. Münchhausen did not suffer from the syndrome which bears his name, nor did he diagnose it, or encounter it in any way. Münchhausen became famous for telling tall tales, and so when this particular condition was discovered more than a century after Münchhausen's death, it was named after him in tribute. Münchhausen syndrome is a psychiatric disorder in which the patient feigns illness in order to gain attention and sympathy. There is a related condition called Münchhausen syndrome by proxy, in which a parent pretends that their child is sick, or even induces sickness in the child, to gain attention for themselves.

Baron von Münchhausen is believed to have told numerous astonishing stories about his adventures. He served in the Russian army, including two campaigns against the Ottoman Turks, before returning home to Germany. There, he is said to have recounted a whole range of bizarre stories about his adventures abroad, such as claiming that he had travelled to the moon (and met the moon people), fought with bears, flown on a cannonball, and visited an island made of cheese.

His amazing stories were published in 1785 by Rudolf Erich Raspe, with the excellent title 'Baron Münchhausen's Narrative of his Marvellous Travels and Campaigns in Russia'. However, many of the book's tall tales are traditional folk stories, which were around long before the baron, and so it's not entirely clear whether this collection was intended to celebrate the baron, or to damage his reputation. Whatever the truth of the real baron's life, this collection made his name synonymous with fantastical tall tales, and so when Richard Asher first described a syndrome in which patients would invent and simulate illness, he 'respectfully' dedicated it to the baron.

Does organ theft actually happen?

'Ah,' you may be thinking, 'this old chestnut!'. There is an enduring urban legend about a gang of well-mannered organ thieves, a legend which has been circulating for at least 20 years, presumably as a kind of cultural response to the real-life practice of organ donation becoming more and more common. The story has been passed around email and the internet in many different incarnations, and has made its way into newspapers, TV dramas, and horror films. The story generally goes something like this...

A lone traveller is approached by a glamorous stranger, usually while having a drink in an airport lounge. The next thing he knows, he wakes up in a hotel room, in a bathtub filled with ice. On the mirror are written the words, 'Call 911. Do not move', and there is a phone by the side of the bath. When the ambulance crew arrive, they tell him that his kidney has been removed by an organised gang who steal and sell organs on the black market.

This version of the story is clearly false. There are no reports of any kidneys being stolen in the United States, and many of the details are clearly implausible. For one thing, a kidney cannot be implanted into just anyone, there needs to be a precise blood and tissue match between the donor and recipient. Secondly, if an organ is to be transplanted, it usually needs to go into the recipient's body extremely quickly if it is to be viable, and the recipient may even need to be present. Thirdly, this kind of operation is complex and demanding, usually requiring at least five people to perform it, over a period of hours, in a sterile operating room, using various pieces of heavy medical equipment, of a type which couldn't be easily smuggled into a hotel room.

However, the fact that this version of the story is clearly an urban legend doesn't mean that organ theft never happens. The market for organs is a commercial one, with organ brokers charging as much as $150,000 for arranging and managing a transplant. Willing donors in the Third World will sell a kidney for as little as $3,000, perhaps even less – this is more than a year's wage in many parts of the world, and people can usually function perfectly well with only one kidney. Demand for organs consistently outstrips supply – at any given time, there are usually around 100,000 people awaiting a donor in the US, but only 5–8,000 registered donors. People who need an organ are by definition highly motivated.

Thus, it is not entirely surprising to find that it does happen. There are no recorded instances in the United States, at least not yet, but there have been reports from other parts of the world, in particular India, which used to have a successful, legal organ trade before restrictive legislation was passed in 1994. In January 2008, several people were arrested in the Indian city of Gurgaon, for running an organised, large-scale kidney theft ring. According to reports, hundreds of men were lured to an underground facility where they were promised jobs, but then tricked or forced into having a kidney removed, to be transplanted into a wealthy recipient. At the time he was arrested, the alleged ringleader Amit Kumar was said to be trying to flee to Canada.

So, the lesson here seems to be that there's really no need to worry about organ theft gangs trying to drug you in an airport lounge, but if you're offered mysterious, well paid, cash-in-hand work at an underground medical facility in India by men brandishing scalpels, you may want to think twice.

26038322R00107

Printed in Great Britain
by Amazon